W

Bee colony must fly the equivalent of 3½ orbits round the earth to produce 1kg of surplus honey.

Rooftop hives in Winter.

The bee is a cold blooded animal. It needs to take up the temperature of its environment. So can fly in lower temperatures, which is important. It will stay inside to temper below 5°C it will be immobilised for quite a long time at temperatures...

Smoker

Bees head for the dark. If a bee crawls up your sleeve, hold your hand high up in the air as bees always crawl upwards.

As a colony & the temperature falls below 13°C the bees cluster together forming a ball on the combs. The top of the ball will be covered with the skins of honeycomb bees where the combs are empty the inside will crawl into the cells making one almost solid ball. The effect of the cluster is to reduce the heat lost from the bees. The bees in the middle eat the honey & generate heat which will warm the others.

Workers make wax in special wax pockets on the underside of their abdomens. This is used for building and sealing the cells.

wax pockets

BEE KEEPING

B

The female worker bee create a pheromone at the start of the season to entice the Queen to lay in it.

Virgin Queens take their mating flight 5-6 days after birth. She flys high in the air followed by many but she too on when the drone has exhausted its mated genetics. The Queen carries a lifetime of sperm in her ovaries.

The Queen becomes infertile after about 3 years. You can tell when the slowly pepper pattern laying random cells. The females have the workers, kill the infertile Queen.

Hives in Spring are placed on the edges of woodlands.

Capture Queen from adult & Queen will be number of bees by two rather tart 2 laid by the honey proximate cage. She can selfishly life starts best.

O

FOR HONEY

Tate Modern 2008 Turbine Post installation. A London Black cab was refurbished. It showed bee stories on its exterior.

TATE ETC

There are bee hives on top of the Tate.

The pheasant was bought from Dazotte in somewhere. Steve & crossed & for me out of the bread. I was very impressed. So we drove to London from Derbyshire and each enjoyed feather as a reminder of a wonderful, interesting...

S

TATE & LYLE

Box Set
of 6 monofloral
beurge candles
honey soap
£28.00

Rooftop beekeeping...

THE ABC and XYZ
BEE CULT...

THE URBAN BEEKEEPER

THE URBAN BEEKEEPER

How to Keep Bees in the City

STEVE BENBOW

■ SQUARE PEG

Published by Square Peg 2012

2 4 6 8 10 9 7 5 3 1

Copyright © Steve Benbow

The Author has asserted his rights under the Copyright, Designs
and Patents Act 1988 to be identified as the author of this work

First published in Great Britain in 2012 by
Square Peg
Random House
SW1V 2SA

www.randomhouse.co.uk

Addresses for companies within The Random House Group Limited can be found at:
www.randomhouse.co.uk/offices.htm

The Random House Group Limited Reg. No. 954009

A CIP catalogue record for this book
is available from the British Library

ISBN 978 0 22 408689 9

All images copyright Steve Benbow except set 1: p2 Jill Mead/Steve J. Benbow;
p4 *top* Jill Mead; p5 *top* Jim Mann. Set 2, p1 *bottom* Eric Tourneret;
p3 *bottom* Julius Domoney; p4 *top* Julius Domoney; p5 *bottom* Eric Tourneret;
p6 *bottom left* Julian Winslow; p8 *top* Eric Tourneret.

The Random House Group Limited supports The Forest Stewardship Council (FSC®),
the leading international forest certification organisation. Our books carrying the FSC label
are printed on FSC® certified paper. FSC is the only forest certification scheme endorsed by
the leading environmental organisations, including Greenpeace. Our paper procurement
policy can be found at www.randomhouse.co.uk/environment

Designed and set by seagulls.net

Printed and bound in Great Britain

For Ned

CONTENTS

Introduction   xi

JANUARY **HIBERNATION TIME**   1
Urban mission • What the bees are doing • New beekeepers •
Individual choices of equipment • Equipment for the first-time
beekeeper • Get planning – with a piece of cake • Struffoli • Oxalic
acid hive treatment for Varroa • How are my bees? • Watch out for
stingers • On the move

FEBRUARY **ESCAPING THE DARK SEASON**   23
Bee obsession • On the way to W1 • Advantages of cold weather •
Be prepared but be patient • A new beginning but a sad parting
• It's all about the cake • Jayne's Honey Cake • Shropshire folk •
Checking over and making up equipment • Peter, Zen master
• Smelling sweet • An old hand • Renovating the old site • Keep
your hives off the ground • A new generation • Getting children
started • A ray of sunshine

MARCH **COMING BACK TO LIFE**   45
Making sugar syrup • Back with my little darlings • What you
need to wear • Varroa and disease • Website for bee disease •
Basic hygiene • Don't look up • Dr Bernays' Honey Gar • On the
road again

APRIL **BACK TO URBAN SPLENDOUR**   65

Moving and positioning bees • Transporting hives • Feeding your bees • How have my bees fared? • Over-wintered queens • Life in the balance • Pollen: a super-food • Propolis • Gourmet honey • Comb Honey with Munster Cheese Foam • Am I ready? • Honey boxes and queen excluders • Making sure all is well • How to do a basic brood inspection • Queen cells • Queens and breeding • Different strains of bee • Smoking your bees • From spring to summer

MAY **MANIC MAY**   95

Spitalfields Market • Áine and Esther • The hazards of swarming • Why do bees swarm? • Scooping up errant bees • Checking hives for signs of swarming • Bigger hives • Artificial swarming • Keep careful records • Still stuck in traffic • The first honey • Tom Bean's Honey with Mint and Peas • We've survived

JUNE **PROGRAMMING THE SAT NAV**   117

Observing how bees navigate • Persuading people to like bees • The old pumping station • Tasting different varieties • Summertime occupations • Summer Shortbread • Bumblebees and wasps • Fetching new bees • Collecting a nucleus • A father to my bees • Worst places to be stung • Major transportation • Thinking locally for varied flavours • Postponing the Hackney move

JULY **COLOSSAL NECTAR AND HONEY FLOWS**   137

Water for bees • Colossal nectar and honey flows • Extra frames in a hurry • Frames or brace combs • Lots of lovely honey • Marauding wasps • Double Honey Ice Cream • Oilseed rape • Mandana • David's London • Honeydew • The Tate gallery sites • Swarm in East London • Zambia

AUGUST **EXTRACTING THE HONEY**   165

Yorkshire heather • Neanderthal man • Shropshire heather • Shrewsbury Show • Class 72 Honey Biscuits • Taking off the honey boxes • Removing the bees • Naked honey extraction • Sugar-shake testing for Varroa • Ditching the drones • Jars for Fortnum's • How much honey to expect • Piccadilly porches • Undercover at the London Honey Festival

SEPTEMBER **THE SEASON DRAWS TO A CLOSE** 193
Paradise in Crete • The end of the season • Combining weaker colonies • Searching for sweetness • A new home at last • Honey in jars or combs • Wax moths • Going corporate? • Upstairs Downstairs • Baklava French Toast • Packing and labelling • Flogging your wares • My golden rules for selling at market • The bee cab

OCTOBER **THE LONG MYND** 217
Fabulous fungi • Lara's Heather Honey Harvest Cake • Leaving the moor • The hazards of high-rise life • Stinky ivy • Getting ready for winter · Rodents in your honey boxes • Beekeeping courses • Innovative designs • Honey shows • New office, new deal

NOVEMBER **WINTER WORRIES** 237
Packaging honey • Open studio • Making deliveries • Senseless vandalism • Animal vandals • Snug at eight storeys • A restaurant delivery • Spiced London Honey Dressing • Cosy clusters • Sustainable food

DECEMBER **TIME TO RECUPERATE** 255
Occasional site checks • Relaxation and reflection • Pre-Christmas mania • Christmas markets • Old-fashioned attention to detail • London Bee Summit • Knocking-off time • Kingy's Hot Toddy • A traditional wassail • Time for new ideas

Thank you 273
About the urban beekeeper 275
Acknowledgements 277
Index 279

INTRODUCTION

I love everything about bees. Their agile yet languorous flight, their barely audible hum (the snoozy backbeat to a baking summer's day) and, of course, their sweet, multi-hued honey. It's still a source of wonder to me that, within a short distance, its flavour can mutate from full-bodied chestnut to sweet lime to silky rose.

Some people might hardly notice these little creatures, but they are everything to me. I adore their varied personalities, admire their magnificent work ethic – an average hiveful flies to the moon and back each year – and am in awe of their humility. Without their pollination, human life would falter.

My passion for bees was fed by an earlier childhood devotion to insects. I believe it started in a coal bunker, pink, concrete and tucked into the back of my parents' Shropshire semi. My secret lair was dark, dry and almost inaccessible to anyone bigger than a skinny little boy in tartan flares with a flash torch, and I was delighted to discover it was also a favourite haunt of the giant house spider.

Row upon row of Robinson's jam jars – the glass magnifying the gangly monsters inside – were devoted to my collection. I punched crude breathing holes into their metal lids with nails. Each day the spiders were fingered and examined, but with near reverence – never harmed, always eventually released.

As my fanaticism for crawly critters grew, my family was encouraged to consume more cosmic-coloured jam, and soon a similar collection was housed on the radiator cover at Long Meadow Junior School. It included winged bees found dead in the school grounds, speared with pins onto bowing cardboard cereal packets. Alongside them sat my collection of cracked bird's eggs wrapped in cotton wool.

This classroom was also the location for my first bee sting – and no beekeeper ever forgets that. Thinking back, I'm certain it was the summer of 1976, the year the country snoozed under a heavy, 20-tog duvet of heat. I was trying to rescue an increasingly distressed honeybee as it struggled to escape through the huge glass windows. The buzzing was a frantic siren call and I'd risen from my tiny chair to help.

All the windows were open, but the bee still failed to find its way out. As I cupped my hands around its frantic form, ushering it towards the sunshine and air – I'd not yet perfected the glass-and-paper routine – it delivered its coup de grâce.

I can truthfully say that the sting didn't really trouble me. What I was most concerned about was the bee's death as its abdomen was torn apart. I didn't react to its injecting of venom into the palm of my hand – this seemed more fascinating than sadistic – and the pain passed quickly. What I didn't then realise was that this was a good omen for a future career – as well as a new member for my cereal-packet pinboard club.

Even at eight years old I'd known it was a honeybee. My maternal grandparents had been keeping bees on their Shropshire apple orchard ever since a government initiative to combat sugar shortages in the Second World War. Granddad's honey was smooth, dark and sweet, and I loved it on a doorstep of white bloomer.

I'd like to claim that my beekeeping skills were lovingly passed down the generations, but there were no cosy fireside chats with this giant of a blacksmith about his mesmerising craft. In fact, the only wise words on the subject I can recall from him were, 'You should always tell your bees your worries.' It's OK, Granddad, I still do.

My paternal grandmother, Grandma Kate, was a formidable midwife who lived on the other side of the county and was also a beekeeper. She was the one who really taught me about bees to start with and a fantastic huge photograph of her and her bees taken in the 1930s is one of my most treasured possessions. I guess she awoke the passion that was lurking dormant somewhere in my DNA.

Over the years my fascination with bees just grew and developed, until one day I woke up to found myself in the strange but wonderful occupation of a full-time beekeeper. This book tells of my adventures along the way, and particularly follows my progress over the year I decided to bring my bees back to London. What I like to call my 'capital takeover'.

By making each chapter a month, I'll show how the beekeeper's year pans out – the dos and don'ts, the little secrets, the big mistakes, and the tremendous rewards. I hope my adventures will inspire you to find out more about these marvellous little creatures. Whether you're in the deepest rural shires, heather-clad moors, windswept sea fronts or cluttered city streets (and I've done them all), the essential skills of beekeeping remain the same.

I guarantee that you will never do anything else in your life which connects you so closely to the changing seasons and the vagaries of nature. Or anything remotely so rewarding and life-enhancing.

JANUARY

HIBERNATION TIME

In days gone by, the start of the year would hardly have triggered mass outbreaks of enthusiasm, energy or motivation. As a commercial beekeeper, I would use the post-Christmas lull, not as an opportunity to implement my New Year's resolutions and join a gym, but instead for the expansion of my waistline; and for recuperation from having worked flat out to a point of apocalyptic bee exhaustion.

Requests from delis and shops for beautifully packaged urns of lightly coloured glowing honeycombs would have begun to arrive at the start of November and continued all the way until late Christmas Eve. By knocking-off time my fingers would be raw from tying intricate bows, whilst my spirits suffered from perpetual festive blues.

Selling substantial quantities of honey stock pre-Christmas would mean my finances were often healthy at the start of January, allowing for a comforting level of reassurance that there would be money in the bank during the lean months to follow – until my dedicated and devoted bees emerged from their own slumber, hopefully in fine fettle for a new season's production in the spring.

Loaded with tins of discounted Christmas biscuits, cakes and chocolates, it used to be that I would hole up for the month in my tiny Butler's Cottage on the sprawling country estate just outside Shrewsbury – once home to Barbara Cartland. I would light large toasty fires and watch my toes glow red through the holes in the ends of my old woollen socks, whilst swotting up on bee-mastering techniques and intricate manipulations from my growing bee-book library, ready to be implemented when the weather warmed.

There was not actually hands-on bee work to be done just yet and, as I became fused to my ageing green velour Chesterfield, my nomadic lifestyle would get put on temporary hold. My two enormous grey cats, Dr Evadne Hinge and Dame Hilda Bracket, would applaud my loafing tendencies.

The previous year would have been spent travelling from one town to the next in my bee truck, my tent and sleeping roll stored in the trunk over my cab, along with a small stove and a few tins of essentials, including loo roll. Washing in streams, drying clothes under hand dryers in service stations, eating roadkill and surviving from the land as much as possible, I'd often long for a comfy bed and a hot steamy bath.

Life is a bit different now that I'm operating on a bigger scale. This year, as soon as the holidays are over, the numerous restaurants and delis who I supply are keen to restock – the Christmas bonanza has stripped their shelves of supplies. This gives me an added incentive to extract the last of my honey and get a welcome boost to my finances. Extraction is rarely a January job. At this time of year, it's far too cold to keep the honey runny naturally, so most sensible beekeepers will have done it months ago, on a warm day in late summer or early autumn. Or they will have done it little by little as it ripened over the season.

Somehow in the rush I didn't get around to extracting my lime honey last year, so on New Year's Eve, traditional night of merriment for the rest of the population, I was stripped to the waist, working through the night. It meant I'd had to stand up a girlfriend, but what else is a bee man to do? Extract his honey and inadvertently extract himself from yet another relationship.

I'd made the place as warm as I could and a few days earlier stood my honey boxes under the hottest light bulbs I could find, in an attempt to soften the wax and loosen the thick lime honey behind it. Then with only a bottle of dark rum for company, I worked till 4.00 a.m. before collapsing in a sticky heap.

When I finally stirred the next morning, it was incredibly satisfying to see dozens of brand new bright orange builder's buckets filled to the brim with glorious unfiltered honey from London's finest lime trees. A truly joyous start to my year.

Urban mission

By two weeks into January, I'm already itching for the bee season to begin. This year there are enormous plans to be implemented. Very soon my bee truck will be chewing up the tarmac once again, but on a different kind of adventure, an urban mission. My bees will need to be in peak condition, as they will be heading south, and this is no holiday.

For the past five years I have managed bees commercially in the heart of Shropshire and West Wales, alongside a dear old friend, David Wainwright. We have had some amazing bee adventures together and shared some rather special moments of friendship in extreme circumstances. Some bee projects have been massive success stories, yielding barrels of honey and prosperity, with bees thriving in wonderful surroundings.

There have also been colossal disasters, with bees at times becoming sick and starving. It has been an emotional roller-coaster but I would not have changed this experience in any way. David is an unsung hero, a true bee-master of humble-ness and wisdom, and without him I would not be in the position I am now. Although at times bee-keeping may sound romantic, more often than not it has been more like unarmed combat, trying to cope with bad weather and disease, despite our unquestionable devotion to and love of the bees.

This new venture of mine will signal a break in our part-nership. It is a chance for me to implement my vision to make London self-sufficient in honey and see if it is possible to obtain honey from all thirty-three of its boroughs – it could be a crazed pipe dream or an inspirational mission. There have been many doubters, who years ago thought that even my original idea of keeping bees on city rooftops was wild.

For this will not be my first foray into urban beekeeping. It will, however, be the first on such a mammoth scale – by anyone, for that matter. Fifteen years ago I decided I wanted to bring a little bit of the countryside to our ex-council flat in Bermondsey, and I installed a beehive full of Gloucester-shire bees behind the lift shaft on the then unused rooftop. The bees thrived and produced huge amounts of dark thick honey in their first year alone.

My bee knowledge back then was good but I still chose to attend a brief course at an adult education centre. Growing up in Shropshire, I had been taught beekeeping by friends and family, but had been apprehensive about keeping bees in such a central urban environment. I need not have worried. They flourished and did not, as I had feared, hassle inhabi-tants or commuters.

For one final time now, I visit David to talk about my plans to head to the capital. He lives in a secluded valley in West Wales, in the shack which he built for himself – the most heavenly place, even if it does have a midge-infested outside long-drop bog, requiring some stealth bottom positioning to avoid being eaten alive.

Surrounded by bees and soft fruit trees, it's a smashing and idyllic haven for bee research and queen breeding and from which various projects are run in association with Bangor University. Its isolation is perfect for all the valley's inhabitants from otters to pine martens, and after years of bee toil and graft, David lives in beautiful surroundings, with 1,000 litres of cider which we made together last autumn, and thousands and thousands of bees.

Each colony has its own locally produced British-bred queen, all named after Welsh princesses. David's favourite is Anwyn 29 – a dark slender queen who has produced an alarming number of offspring – now in semi-retirement in a beautifully crafted bee shed that captures the tiny amount of sunlight the valley has to offer during the winter.

Our meeting does not go well. David and I have always managed the colonies together – he in West Wales and I in Shropshire – and I feel guilty about leaving him as managing bees in large numbers by oneself is seriously challenging.

David is a man of few words but his facial expressions say it all. He's partly hurt and I'm sure nervous of the future for his own bee ventures alone.

My feelings are mixed too. This part of the world is where I was first introduced to the mesmerising craft of beekeeping and I will be leaving behind many memories. My grandparents on both sides of the county were my first inspiration – the giant photograph of Grandma Kate tending her bees will one day hang in my new studio. And there was another silver beekeeper who kept an actual bee village on her Lavender Farm near Bridgnorth. Each hive stood behind a hand-painted façade of a church, cottage, bank or school. When the bees swarmed they would hang from miniature street-lights and signposts. It was magical. We'd consume gallons of honey-sweetened black tea in her rustic kitchen, while discussing life in Hive Village. I had met my first bee guru.

But I know that London is the best possible choice for the health of bees in the future and for the quality of their honey – also perhaps for my own sanity as I worry about the changes happening across more and more of our countryside. I already keep some elaborate and ornate hives on top of Fortnum & Mason's and am keen to expand across the capital. The famous department store came to me last year asking if I would install four hives. Since their arrival, successful bee tours have taken place and there are two live webcams through which you can follow the progress of the bees online.

During recent student demonstrations, protestors broke into the store and occupied a section of the rooftop. Friends, having seen it on the news, phoned me in a panic about the bees and I was able to go online and check that the beehives were still there ... which they were, the bees apparently unaware of the ensuing commotion. These same friends have been pushing for a naked online streak but those days are behind me and the risks are too high – besides, I wouldn't fancy the stings.

Out in the more concreted mono-floral parts of the coun-tryside, bees face the constant threat of modern pesticides, particularly those used on oilseed rape, a crop that does far more damage to bees than most people realise. By compari-son, 42 per cent of London in open space and a further 24 per cent is covered in private gardens. In addition there are railway sidings and green rooftops, so these figures could be even greater, and people are always planting up window boxes and replacing park bedding. With such varieties of pollen, the honey also tastes fantastic. Though the bees out in rural West Wales and Shropshire are not so threatened for the moment, I feel I owe it to future generations of bees to develop city colonies.

An amazing French photographer, who recently visited me for a new photographic book he is compiling on global beekeeping, told me of commercial bee farmers in Berlin, who move beehives into the city specifically for the lime harvest: the city has a huge number of mature lime trees as well as extensive woodland generally. The operation is all very covert, to avoid detection and possible panic, with hundreds of hives being moved into position for this specific crop. I had no idea this happened and it gives me confidence about my own operation. Apparently each hive moved into Berlin during the tree's short flowering period produces on average 25 kilos of thick greenish honey. No wonder the lime is highly prized.

We could do with a few more mature trees in London, lime or otherwise. The increase in urban beekeeping over recent years has forced local bee associations to stir themselves. After an initial panic over the amount of available forage, I'm glad to see they now appear to be working together with local councils to ensure that there is a wider range of nectar-yielding plants in the city. The new wave of guerilla gardeners are also doing their bit. This can only be brilliant news, making London a greener place overall.

Having bees in London is good for the city: not only do urbanites get a chance to enjoy local produce, but it helps to pollinate every green space, from tiny gardens to public parks. So, not only for the well-being and prosperity of the bees, but also for Londoners themselves, I plan to break away from my Shropshire bee shed and put my plan into action.

What the bees are doing

As the month progresses, the first snowdrops appear, signalling that pollen is starting to be available for the bees to forage on. Small clumps of these divine flowers attract the bravest bees out of the hive, waiting their turn to land on the pollen source. When people ask me what they can plant in their gardens to help bees, I usually mention these January bulbs – they are an essential boost for early bee growth.

Apart from these intrepid few, the bees mostly remain inside their hive, grouped together; not exactly sleeping, but huddling in a catatonic state, shivering and vibrating in order to maintain a toasty temperature of 32 degrees. If there is a snowfall, the beekeeper should brush it away from the entrances – although an old boy did once tell me a pile-up of snow could actually keep bees snug in cold weather, providing there is sufficient ventilation reaching them.

Bees can fly in zero temperatures; the problem arises if a bee stops flying and therefore stops using its flight muscles while it's away from the hive. It is unlikely to have the energy to start up again and if the temperature is below 8 degrees, the bee will almost certainly perish.

Snowdrops and later on crocuses will be providing bees with their first source of pollen. This they will bring back to the hive in special pollen sacs on their legs, to deliver a crucial supply of protein for building up the hive's brood. Pollen is essential early in the season for the development and generation of new bees.

At this time of year, it would be too cruel to deprive bees of their pollen – not to mention detrimental to overall hive welfare. But later in the spring, I'll attach pollen traps to the fronts of the hives in order to gather it. For the past year, I've been selling pollen at farmer's markets, where it goes down a storm with health-conscious ladies of a certain age, and sufferers of hay fever who believe it helps their symptoms.

Luckily, there are still a few months to go until I feel the telltale tickle of hay fever. The cold weather keeps pollen at bay, but it also makes inspecting hives a sensitive operation of stealth and care. It's important not to disrupt the bees too much or cool down the hive. From the bees' perspective, opening the roof must feel like someone swiping away a warm duvet on a cold winter's morning, only to throw on a bucket of iced water.

New beekeepers

For the uninitiated, this is the perfect time of year to ask yourself if keeping bees could be for you. Would it fit with your lifestyle? Do you have the space to keep bees, the time to concentrate on them? Start swotting up. You will quickly become engrossed by bee behaviour and marvel at their very existence. This could be a decision that will change your life.

Find out about bee courses. There are plenty of theory courses to enrol onto in January so you can grasp essential bee knowledge now. These are often run by local associations or keen amateurs – but choose wisely and speak to those who have already attended if you can. I got so excited when I first did mine that I was raring to go when the milder weather arrived in the spring, having spent weeks in the classroom of an adult education college in New Cross.

By the end of January, if you do decide to take up beekeeping, you'll need to have touched base with breeders and got in the queue for new colonies (although you won't be collecting these bees until early summer). It's always good to look for reviews and other customers' comments when searching for a new set of bees. In the past, I've collected swarms to restock my bee supplies, but as their provenance can often be unknown, I now prefer to order new colonies from reputable breeders. You may prefer to fetch them yourself, but remember that bees can be sent in the post too.

Individual choices of equipment

In January, the established beekeeper will need to clean up old equipment ready to be used again. One of the jobs that this involves is boiling up old frames in washing soda to sterilise them, before steam washing them with a power hose.

It is also a good time of year to start thinking about what equipment you will need for the year ahead, especially as manufacturers offer discounts in the quiet season. Beware of taking on too much while you are still learning the ropes. Assembling a flat-pack hive might sound like a simple mission, but if you live in a city flat with limited space, you'll need to think about where you are going to do it and ideally recruit some help.

In my early years as a beekeeper, I built hives sometimes on my rooftop, other times in my council-rented garage and occasionally, if I was pushed for space, in my communal stairway, which was often frequented by crack users – taking what they had scored in the basement's den.

Equipment can be bought ready assembled, although it is more costly. I buy some of mine from Italy where it's cheaper and then I make smaller items myself, such as hive roofs, floors and mats to keep the bees cosy.

Beware of suppliers who inflate prices of ordinary items that are widely available from hardware stores. For example, a honey bucket can be bought more affordably when it's just called a bucket. Deeming everything 'bee this' and 'bee that' strikes me as a wheeze to hoick up prices. I often get these things cheaply from a builder's merchant, though this does mean I have to sterilise them with a light disinfectant first.

Having said that, sometimes the best tool is a specially designed one. For instance, do get a proper hive tool rather than using a screwdriver like the one used by a pioneering urban beekeeper who I met in Paris. Jean Paucton is an elderly character, who tends his bees on the roof of the Garnier Opera House. He leaps around the gargoyles like a youth and produces a light delicate honey that sells in the gift shop for a fortune. Though he manages with a screwdriver, you might find that a properly designed hive tool with a hook on the end and a straight flat edge will serve you better.

In many ways, choosing a hive is one of the biggest decisions that you might make at the start of the year, as you plan for the months ahead. I would recommend hives made from cedar wood, which is less susceptible to bug infiltration and rot, and needs little maintenance or wood treatment – its natural oils help preserve the wood. I have some cedar hives that are over sixty years old.

To make beekeeping more affordable and accessible, hives are now widely available in Douglas fir or other common softwoods, but these will definitely require some form of treatment and protection from the elements.

Many people decide to paint their hives dull colours, but I prefer bright shades, so they look more like Mediterranean hives. You can always spot my hives as they are the most vibrant ones, with not one single colour matching – and when I get bored painting in blocks I add polka dots, stripes and swirls. I like to think the bees also prefer these jolly colours, not just because they look like flowers, but because they can learn to recognise their own hives. Sadly, however, bright hives are more easily spotted and therefore are more likely to be targeted by vandals.

To prevent hives from rotting, beekeepers used to paint them with creosote and other caustic substances. Fortunately this is no longer acceptable due to the harm these toxic chemicals can cause the bees, but do check your hive paint is bee friendly before applying it.

Equipment for the first-time beekeeper

The must-haves

• First and foremost, a hive.
There are various hives available to buy in the UK – I have tried them all but I love the Modified Dadant, introduced to me by David. I find it works perfectly in urban areas, as it allows me to expand the brood nest in the spring with the use of a dummy board and close it down again in the winter. The frames are large and give the bees plenty of room, which helps reduce swarming.

The majority of my own hives are made of Douglas fir to keep the cost down and then painted with crazy colours, but for those with deeper pockets, you could look at cedar which will easily outlive you and will require little attention against weather and ageing. Each hive will then require a set of three to four supers. Also three to four honey boxes.

Each hive also needs a mesh floor to allow air to circulate around the bees and help with the fight against Varroa;

during treatments, the mites will not be able to clamber back onto the frames and bees and will fall to the ground. These I make myself – from a pattern taught to me by David. He uses pressurised and treated timber, to prevent rotting, and a fine zinc mesh. The floors are made from a series of battens, positioned to allow the maximum amount of air around the bees when they are transported.

The roof is a simple zinc-covered affair and underneath is a hive mat or crown board. Once again I use David's design – foam board from a builder's merchant keeps the bees snug – but most use a simple board that also works as an escape, using small bits of plastic that are inserted to prevent bees from re-entering the honey store during cropping.

- A feeder for syrup.

I talk extensively about these later on. My recommendation is a Miller feeder for rapid deployment of syrup in the spring and autumn, filled with wood wool to prevent the bees from drowning.

- A good smoker.

Choose one with strong leather bellows, and spend a little extra for a cage around the chamber to help prevent you burning yourself. I love to put the smoker between my knees when I work a hive so I have two hands free – or use the hook on the top to hang it from the hive wall – so the cage is an important feature.

- A strong hive tool.

I like the ones with a hook on the end and a straight flat edge.

- A bee brush.

The best thing for gently brushing bees off the combs during cropping is a goose wing but if you can't get one then a commercial bee brush will do. Try getting a goose wing from your local gamekeeper. A wildlife centre in London did once manage to secure me a few when some troublesome Canadian Geese were hassling visitors and they were taken out one evening – I'm not sure what became of the meat but the wings were useful bee brushes.

Other considerations

- A good double-sized white bed sheet.

Use this for running in a swarm or for wrapping up a collected swarm. I always have one to hand as it allows the bees an easy runway into the hive: position it so they can clearly see the entrance. It also makes spotting the queen easier and saves them having to cope with tufts of grass. A sheet around a cardboard box was the way I was taught to collect a swarm – it prevents leaks but still gives a bit of venting.

- Some nucleus boxes.

In time you will need these for making young colonies in the spring. It is important, not just for commercial beekeepers, to maintain hive numbers and keep your stocks young and fresh, and making nucs or rearing your own queens is a good way of maintaining numbers.

- A few queen cages.

For holding young or troublesome queens. Empty match-boxes will do.

- A good note and record book.

For noting down what you've done with each hive, what you need to do next time and what you need to bring with you.

- A shedload of gaffer tape.

For fixing gaps and holes on hives in transit or for even wrapping around your boots to prevent bees creeping in under your suit – its uses in beekeeping are endless.

Get planning – with a piece of cake

January is also a month for planning, a chance to look back at the previous year's exploits and assess what could have been done differently and what could be improved. A beekeeper can expect to spend time working out what he or she wants to achieve from the bees over the following season. My advice is to start a new record book for each hive – essential for individually monitoring each one and for learning from your interventions.

Last year, for instance, the honey flow stopped at the end of June in London and there was no autumn honeydew, which made me realise, in hindsight, that the bees could have had an outing to the heather or wood sage on the south coast. What this profession has taught me, if nothing else, is that no two seasons are the same; it's hard to predict whether a pattern will be repeated the following year. My two nuggets of advice are that you should expect the unexpected and make hay while the sun shines. For wannabe bee farmers you should also get yourself a good osteopath and sign up to online dating.

It's always best to do your planning when fortified with tea and lots of cake. This year I have been given a recipe by Bea Vo, owner of Bea's of Bloomsbury and other fine tea rooms across the capital. Her commercial bakery is in Bermondsey, my old manor and an area where I hope to secure a railway arch studio for my expanding honey business.

This recipe for her twist on Struffoli, a 'favourite fried doughnut treat', works well with London honey.

Struffoli

For the doughnut pieces:
500g plain flour
zest of one orange
zest of one lemon
a pinch of paprika
½ tsp salt
7 eggs – 6 whole, plus 1 extra yolk
1 tbsp dark rum
1 litre vegetable oil, for frying

For the sauce:
500g London honey
juice of half an orange
a good shot of dark rum

icing sugar, for dusting

Mix the flour, orange and lemon zest, paprika and salt. To do it the old traditional way, dump the flour on the counter and create a small well. Fill with eggs and 1 tablespoon rum. Incorporate with hands until you get a nice smooth dough. Wrap in cling film and place in refrigerator for one hour.

Separate the dough into quarters and roll each into a long

rope about 1 inch thick. Cut into small ½-inch pieces like gnocchi.

Heat the oil in pan until it reaches 190°C/350°F/gas 4. Drop balls in a few at a time until they turn nicely golden and puffy. Remove with a slotted spoon and set aside.

In a wide pot, add the honey, orange juice and a shot of rum, and heat over a medium heat until it's quite warm and the honey thin. Add the doughnut pieces to the honey mixture, and stir until well coated. Remove from the heat and let the sauce cool in the pan, stirring constantly to keep the honey coating even.

Pour onto a nice plate (it'll harden as it cools so you can shape it nicely.) Dust icing sugar on top.

Oxalic acid hive treatment for Varroa

The final task that I tackle before the end of the month involves treating hives with oxalic acid to reduce the risk of losing colonies to Varroa, a much-feared parasitic mite that attacks honeybees. It must be done during this dormant month, before the queen starts to lay properly, as the treatment could otherwise kill bee larvae. Oxalic acid changes the PH level within the hive, either killing the mites that cling to the backs of adult bees or causing them to loosen their grip and drop through the mesh floor of the hive. Oxalic acid will only kill mites on the bees and not those on sealed larva, so this is another reason to treat when there is no, or minimal, brood.

To see how bad an infestation you are dealing with, you can put a sticky board under the mesh floor which catches the mites' corpses. A sticky board is simply a white sheet of card or plastic covered in a mixture of olive oil and Vaseline which prevents mites from clambering back onto the bees. It also stops the wind from blowing them away before you have counted what we call 'the drop'.

Regular monitoring of your bees for these mites has to be one of the most important things you must do as a beekeeper as they will kill your colony off if they remain untreated or ignored. This is a grim fact but it's important you realise this early. It is worth noting that your mite numbers will be greater in the autumn when the season starts to draw to a close. It is also the time when you remove the honey boxes and bees spread across several boxes will be condensed into one for the winter.

Every hive needs to be zapped really quickly with a precise dose of diluted acid from a vet's syringe. To do this, crown boards – the ones below the roof that keep the bees snug – are carefully cranked up and a squirt is given to every frame of bees – approx 5 ml per frame. This should involve minimal disturbance of the hive, since it's still winter hibernating season. The weather needs to be cold enough so your bees are in a tight cluster, but not freezing. The treatment is implemented at this time partly because when the bees are tightly clustered, they can all be treated together – it's unlikely that many will be out flying.

This treatment used to be considered maverick but now it has widely become essential and it is seen as the most effective method of battling Varroa – so long as you proceed carefully when applying the acid. In the past, intense chemical insecticides were more commonly used; but Varroa mites have mostly become resistant to them, thanks to their overuse both in treatment and in the fields, largely those planted up with my foe, oilseed rape.

How are my bees? In the second week of January I open up the hives on the roof of Fortnum's and at first I am alarmed. There is no sign of life and I think all the bees have died. Slowly, however, they stumble out from where they were holed up at the bottoms of the frames – sorry, Bees, this is truly a rude awakening.

When I come to hefting my other beehives – to check they're not too light which would indicate that the bees are starving – I'm alarmed to see large numbers of dead bees in front of some hives, which suggests I've been badly hit by Varroa this season. I need to remove the corpses from the entrance blocks to ensure that any flying bees can still get out of the hive on warmer days to find water and early pollen. I take time to do this carefully so as not to disturb the cluster of catatonic bees.

Disturbing though this discovery is, the fate of the bees is no longer in my hands. All I can do is hope that they will pull through. I've done what I can to ensure their well-being – they are largely on their own now. There will be casualties before the warm weather arrives, that much I know for sure. My only hope is that these are not huge and that it is nothing major that will slow their development during the spring.

Watch out for stingers A common problem for beekeepers at this time of year is discovering that some of their bees, attracted by body heat, have landed somewhere on their person, lodged themselves in an unusual spot, and then failed to summon the energy to take off again. Chances are this will only be noticed when the beekeeper has reached the warm sanctuary of a vehicle or centrally heated house, where the bees will suddenly re-energise – usually with a vengeance. That's why you need to take great care when removing smocks and veils.

My advice to first-timers is to work in a buddy system if possible. Before returning indoors, give each other a twirl to check for waifs and strays – and help prevent an early encounter with a stinger. David and I arrived in a Chinese takeaway last year after an early-season bee day and spent the next ten minutes, having already ordered our supper, catching from our overalls the bees which were warmed by the lights. Apologies to the owners of the China Garden take-away, Machynlleth – it must have been alarming.

On the move In order to offer comb honey to customers around the year, I store my prized combs in cold storage, where each box is wrapped in cardboard to help minimise condensation. Reassured that these combs are in tip-top condition, towards the end of the month I get them wrapped and despatched to David's factory in West Wales, where they will be cut into chunks and then packaged for delis. This logistical exercise is something I want to change this year, and so begins the search for a suitable studio and office space in south-east London.

Up till now I have thought nothing of sending my London honey off to David's factory for processing. Each box is carefully wrapped in a black bin bag and interlocked with the next. The complete pallet weighs 500 kilos and I wrap it with industrial shrink-wrap to make it one big unit.

Throughout the year, hundreds of honey boxes have been transported around the country from the various crops our bees have produced. Only one courier company has ever complained and that was because a few bees were still attached to the honey boxes.

This time, my worst fears come true. The pallet of London honey arrives at David's factory in Aberystwyth completely smashed. He phones to break the news with his most calming voice – he knows that this will be a devastating blow to me and my business. David can show great compassion at such times along with a straightforwardness I applaud.

The driver managed to mangle this precious load in the final few metres of its journey by slamming his brakes on at the last minute. It turns out the load fell off the truck crashing to the deck, as it hadn't been adequately strapped down. I receive photos via email from Matt, a good friend there – he's almost in tears. For my honey to be wrecked at this late stage is not only a personal disaster, it's a great disservice to the bees that have laboured over this precious crop. I have let them down by entrusting it to some idiot, who has no idea of the impact of his actions.

Although some of the honey can be salvaged by pressing the combs, the combs themselves are ruined, along with this year's plan to sell it for a good profit. The load was insured, but not for nearly enough – barely a fifth of the overall price it would have made me.

Over the following days, I reach an all-time low. I was relying on those funds to purchase much-needed equipment for expansion and to bolster my big move down south. It's hard to keep up my spirits, even though there is good news in other areas. My beginners' beekeeping courses, for example, are packed out with keen new amateurs wanting to learn the ropes. I notice a renaissance in Londoners signing up to this hobby as part of a New Year new look, and there's an increase in volunteers keen to help bash any possible new sites into shape across the city. New super-sites are an issue but I have one hope – an old bee-master in the north of London, who has contacted me out of the blue.

It's great news for my urban mission, but sadly nothing can take my mind off the fact that my business has taken a hammering – all thanks to the incompetence of one moronic van driver. It's not a good start to the year but the month closes with David offering to give me some offspring from his Welsh queens when the weather warms – an amazing gesture that means my bee stock will be fresh and young when the season starts. I just need to find some boxes to house them in. Not even wild horses could stop me now – I'm on the move.

What potential new beekeepers should be doing:

- Think about whether you could you keep honeybees. Is this craft something that could be workable with your lifestyle, space and time?
- If so, enrol on a local theoretical bee course in January – they become booked up quickly at the start of the year.

More extensive beekeeping tips:

- Consider any new equipment you might need for the year and set about buying and making up items to save money and to be ready for the season.
- Clean up and sterilise existing equipment for reuse.
- Check your hives for levels of Varroa and treat with oxalic acid.
- Plan your strategy for the year: make a list of how you might wish to manage your bees, i.e. expansion of your stock or new locations.

FEBRUARY

ESCAPING THE DARK SEASON

Confession time. I've always disliked February. Grim dour February. It's not that I suffer from Seasonal Affective Disorder but it's a close-run thing. Winter's dragging. Drizzle, grey clouds and low temperatures are the bleak norm; fresh sunshine – even with an icy refreshing winter chill – an exotic stranger.

Before becoming a commercial bee farmer, I had been living in London and working as a travel and documentary photographer. It was an extreme lifestyle that saw me jetting off to some fabulous places, sometimes only for a few hours, before picking up a tropical bug and setting sail for a new location. I was known in person at the then Hospital for Tropical Diseases at the back of King's Cross. Much of it was fun, intoxicating even, covering stories that saw me surviving on a tropical island, rambling naked in the Californian desert and acting in Bollywood movies.

In February I would often try and shoot features in warmer climes, anything to avoid the short days and relentless gloom. In the distant past I even once took the whole of this month off to teach sailing in Malaysia – and get an exhilarating, life-enhancing blast of tropical heat.

Bee obsession Glamorous though this lifestyle may sound, its appeal didn't last. As my professional interest began to pall, so bees began to creep in. I started tacking extra time onto trips to visit urban beekeepers at random points around the globe. In the morning I'd do a shoot, and then with my work duties out of the way, I'd slope off to meet some mystical character I had tracked down on the Internet.

Picture editors were initially tolerant of my new obsession, despite work arriving late, punctuated with random images of grinning bearded beekeepers – a few were undoubtedly sweetened by weeping jars of aromatic honey. I even kept a pair of pink rubber gloves in the bottom of my camera bag, carefully folded underneath a black bee veil that I could slot over my panama hat if I stumbled across some unexpected hives.

From the one hive tucked behind the lift shaft of our ex-council flat in Bermondsey, within only a few years that had expanded to over fifty hives scattered across the capital in some amazing locations, such as electricity substations and wildlife parks. I was still having to make a living as a photographer but was getting closer to the day when I would give it all up.

Back in those first years, my bees were kept covert and out of sight, deliberately to attract little publicity. To avoid detection, I would tend them during the early hours, slipping quietly around them like a cat burglar, for some were on sensitive sites, whilst others were squatting and on borrowed time until the developers should move in.

An apiary's position could have been blown if the bees had decided to swarm or if there had been a particularly frantic honey flow. Although it is rare for people in the city to look up, if they'd glanced skywards on a warm day, they might have witnessed a frenzy of activity and clouds of diligence as my bees worked the untapped resources of urban nectar on offer. There was something incredibly serene about being seven storeys up in the heart of a major city working with such an ordered society, while below were stressed commuters, congestion and mayhem.

These were exciting and exhilarating times. The bees were not just surviving in the city; they were prospering. Yes, my neighbours had at first been alarmed by seeing bees bouncing off the windows as they were buffeted by winds on the journey to their fourth-floor hive, but as the illegal rooftop garden was developed around the bees the residents benefited too – not only from the splendour of the garden where I grew tomatoes, potatoes and salad along with bulbs, shrubs and vines – but also from its honey harvest. I soon realised that jars of golden loveliness were the way to bring an end to people's suspicions.

Now that I was reunited with my passion for small furry critters, friends suggested it could be an interesting business idea for me to service bees for companies around the city. I had thought about this years earlier but had concerns over safety and the welfare of the bees in a harsh environment. My fears were quashed as I developed my beekeeping skills and expanded my bee business, which saw my bees producing a rather unusual but eclectic-tasting honey that I soon discovered varied considerably from one site to the next.

Forming what was to become the London Honey Company – except back then we called it Beesplease and we had a slogan, 'Nectar for urban workers' – I set about promoting the good work of bees in the capital. Educating people that these pollinators were integral to such a dynamic and green city was no easy task, as the mere mention of bees to some would trigger utter panic.

So I decided to take a research trip to New York where I had heard of a maverick beekeeper called David Graves, who kept his bees high above the streets of Manhattan. He paraded around at fifteen storeys' height without a safety harness or a veil and used an old T-shirt to brush bees from combs he was cropping. He moved both the bees and honey on the subway, wrapped in bin bags – this not only prevented robbing by the bees but also contained any sticky leaks.

It's a procedure I still follow today when I am taking honey from a particularly urban site, as it also prevents any bees from flying at office staff as I ferry honey from their roof. It got me in hot water with the council in the early years, who were convinced that because I was moving large quantities of bin bags in and out of the flat at strange hours I must be a drug dealer. They had failed to notice the crack den that had been in the basement flat for years.

In New York, it was also a rather clandestine affair, as in those days bees were classed as wild animals and forbidden from being kept in the metropolis. That was the official line anyway. David Graves managed, however, to demonstrate not only that keeping bees commercially in an urban environment was feasible but also that the honey they produced was exquisite – even if it was alarming to hear that his bees would always die over the winter due to the severe temperatures the Manhattan rooftops were subjected to. He replaced the bees every spring.

By this stage bees were taking up more and more of my time, so five years ago I decided to give up photography altogether and become a full-time commercial bee farmer. It made sense to go back to my beloved Shropshire where I could immerse myself in beekeeping and expand my knowledge through close work with David, while at the same time breathing some fantastic air. I also hoped to give my young son Ned the sort of country upbringing I'd had.

As things turned out I ended up on my own in the cottage during the week and at weekends I would race back south to London so I could spend time with Ned and his mother, and see to my city bees and their requirements before the temperature dropped in the evening. Then it was back to the Vale and another week's toil. I was glad that the farm where I had my bee shed was such a close-knit community.

On the way to W1

A very well-spoken man from Fortnum & Mason had been the first to contact me to see how feasible it would be to have bees on the roof of their store. As it was being renovated at the time a journey was planned for their ornately commissioned hives to travel around the UK until they could be settled on the rooftop in Piccadilly, and the first stop was a country estate in Shropshire – how convenient.

In an auction some years ago I found a book called *A Jar of Honey from Mount Hybla* and dated 1859. At the start of Chapter One there is a reference to passing the window of Messrs Fortnum & Mason and beholding a little blue jar of Sicilian honey in their window display.

Fortnum's have always prided themselves on their honey and prior to my appointment as their 'bee-master' I had always admired their gloriously stocked shelves, exhibiting great urns of comb and exotic honeys from such places as the Pitcairn Islands.

I'm not sure the bees appreciated their new palatial residences but they produced an excellent honey that year from the nearby heather moors and were the focal point of a display garden at the Chelsea Flower Show a year later. Then the bee roadshow travelled to a glorious cottage garden in Henley, before joining the newly installed air-conditioning vents and ducts on the store's somewhat industrial rooftop in Piccadilly.

The smell of breakfast from the various restaurants downstairs would drive me to distraction during morning visits to service their hives. This wasn't helped by the fact that I would probably check on the bees more than was necessary, as I was paranoid that they would swarm onto the adjacent Cavendish Hotel. In fact it was the wild wind that would be the biggest challenge for the bees in those early weeks as it funnelled down Jermyn Street, and I erected windbreaks made of willow withies to aid landings and take-offs.

Advantages of cold weather

There's even a potential silver lining to the dark winter clouds in Blighty this year. The season has been so severe it could well have killed off or subdued any disease amongst the bees – at least that's what some old boys tell me. I tend to agree. It has always been accepted farm wisdom that harsh winter weather hammers disease amongst livestock and crops. Perfect.

An added bonus is that the brutal temperatures should keep the bees inert. If they remain tightly clustered, using minimal energy until consistently mild weather arrives, they're less likely to starve and more likely to survive. Why's that? Well, activity means energy – and that means eating. Premature mild conditions encourage bustling bees when there's still the chance of a cold snap. Ideally you still want most of the bee ball hunkered down until a significant change in the season and an established rise in temperature.

Be prepared but be patient February is a manic month for the beekeeper, getting ready for the season to begin again, but you need to keep patience too. Anything you haven't already done in January needs to be sorted now as soon you won't have time.

Beginner beekeepers need to put together the equipment already received, and if you haven't yet placed an order for new bees from a breeder then you need to hop to it before the end of the month. Even though your bees won't be ready for collection until June, breeders operate on a first-ordered, first-received system, and they close their books early. Your colony will probably include five or six frames – approximately 10,000 bees – and a new, young queen. This is helpful for the novice beekeeper as she is less likely to want to swarm in her first year, making your first year with any luck a peaceful one.

Do join your local beekeeping association. They can give real support and mentoring – for both the beginner and the established beekeeper – plus a degree of insurance through the British Beekeepers' Association. Associations are generally very active with new members, and some have even been forced to limit membership because of increased popularity. In London I think there is scope for a Central London beekeeping association with all the new beekeepers.

By this stage all you established beekeepers will also be wondering how your bees have fared through the winter. Like me you're probably very keen by now – actually this year, because of the cold, I'm anxious – to see if the weaker colonies have recovered and whether the queens have started brood rearing again. Any peeking would still be hazardous, however, so be patient. Your time will come. These are not pets, remember. Instead occupy yourself by getting ready for when they emerge, and do, of course, make sure the entrances to the hives are clear and not blocked with the bodies of any bees who've died naturally.

Keep an eye out on the warmer days – none so far this year, alas – to spot any adventurers who may be out foraging on early pollen. If there are bees coming back with their pollen sacs full, you know that the queen is laying, since pollen is food for grubs. If you've got the time, you might even want to mix up sugar syrup so that you're ready for the start of spring feeding in mid to late March when the weather warms – as soon as the bees become active it's possible they will need a boost. In other words, February is the time to be observant and to get ready.

A new beginning but a sad parting

For me this February's different from usual. I'm chipper. In fact, I'm rather excited. Most of the winter is now behind us and I'm itching to implement the major new bee strategy I've been developing for the last six months. I'm particularly keen to reveal the fruits of my incompetent woodwork. Yes, I've knocked up some new hives and equipment and covered them in garish paint – ready for my city slickers.

The companion I will miss the most when I make the move back to south-east London, aside from my ageing Roberts radio, will be a bow-legged Jack Russell called Herbert. Officially he belongs to Joe and Ann whose farm, only two miles from my Shropshire cottage, is where I have my bee shed. Herbert shares my passion for anything sweet, baked and wrapped in bright cellophane. For daily cake consumption, two boxes of French fondant fancies is my record. I should be huge, but I'm one of those folk who can never sit still and I always manage to burn it off. I find I eat loads over the winter, as the job is so physically demanding – I also have a sweet tooth and honey does not help.

It's all about the cake

My top five bought cakes are as follows:

- At number 5 is Battenberg.
Its sweet sugary almond paste is a welcome energy rush mid-afternoon, especially during winter frame assembly.

- At number 4 we have jam tarts.
OK, so strictly speaking they're not a cake, but a box can easily be consumed across a day in the field. They are bite sized and can be crammed in whilst driving with safety – an important attribute.

- Number 3 is ginger cake.
It's easy to nibble at and can be moulded into balls and smuggled in through a tiny gap in the zip of the veil when those breaks don't arrive and you need to just crack on with bee work.

- Number 2 has to be an Eccles cake.
Again it's not exactly a cake, and they're often hard to find, but those rich sweet currants make it seem as though there might be some goodness in it. The St John Bakery in Bermondsey does an amazingly gooey version.

- At number 1 we have the Welsh cake.
These can vary considerably but I adore them warm with a little knob of butter. They are my beekeeping tip-top shop-bought cake – although there is a small bakery on the Welsh

borders near Welshpool that makes their own version, a divine flat griddle-like scone, which can almost be classed home-made.

But my favourite cake of them all is my big sister's honey cake. We both learnt to bake from a very early age. Cooking was something that never really interested our mum – sorry, Mum, but true – and my sister and I began by making Christmas dinner. We soon graduated to trying new recipes and Jayne always comes up with some fab cake ideas. I love getting a text from her to say that she's working on a new one; I'll be there in a shot before my nephews squirrel away every last crumb. Her honey cake is rather special, so it seems only fair to share it.

Jayne's Honey Cake

6oz margarine
6oz caster sugar
3 eggs
9oz self-raising flour
1 tsp baking powder
1 tsp cinnamon (sieved with flour and baking powder)
2 tbsp honey (nearly half a jar)

8-inch greased tin and baking parchment base
oven at 180°C/350°F/gas 4

Mix all ingredients with an electric mixer for 1½–2 minutes. Put in tin and bake in centre of oven for 1–1¼ hours.

Shropshire folk Herbert spends much of his time lounging on two shabby Ford Anglia leather car seats, pushed up against the back wall of the former pig-farrowing shed which doubles as my workshop and bee HQ. I'll be sorry to leave him. In fact, I will greatly miss all of the farm's community. They have formed a huge part of my life and been an integral infrastructure that has made the extremities of my life as a bee farmer manageable.

There is Henry Griffiths, the most ingenious fabricator/blacksmith who can construct virtually anything, just as long as it can be made from boxed steel and doesn't require any particular delicacy or refinement. He created the rear deck of my truck, removing the standard back and replacing it with a six-foot tent-like structure. His forklift has been integral to bee movements in Shropshire over the years – loading hives, sugar and full honey boxes. I will have no such luxuries down south and how I'm to deal with the huge weights involved in beekeeping is a real concern.

Mark 'Chippy' Humphreys is the polar opposite to Henry on the farm – he loves wood and old tractors. He generously gave me a key to his workshop last year, which entitled me to some fabulously scary bits of machinery, perfect for those little fiddly bits of hive component repairs and construction. He also reunited me with the old grey Massey Ferguson that belonged to my Uncle Dickie (Dad's brother). It had been standing in Mark's sister's field on a nearby farm for years since my uncle's death and I have great plans for it in the capital.

Then there are Joe and Ann, owners of Herbert and of the tumbling and gorgeous farm where I have my bee shed, and the most delightful and generous couple imaginable. Joe has come across me on numerous occasions around the farm, slumped, exhausted and weary from bee labours – the rent for my shed alone would not cover the amount of tea and cake I have eaten in their parlour over the years.

All of us would congregate for morning coffee at 11 a.m. and then tea at 3 p.m., sitting on newspaper to prevent our overalls from leaving marks on the chairs. World politics, the plight of bees and the *Telegraph* crossword were just some of the matters discussed and dealt with, plus my unruly scraggy beard.

Herbert the Jack Russell would never survive on the mean old streets of London, although his fantastic capacity for ratting might come in handy. Those dewy eyes can be deceptive – really he is a coiled spring waiting to pounce as, trawling through all the winter-stored hives in the larger hay barn, I evict the rats and mice that have been squatting there. Rats can cause real destruction, gnawing their way right through the boxes, while the mice can't resist the lingering smell of honey and beeswax and build their nests amongst the frames. I should have put everything away more carefully in the autumn, sealed tightly against rodent attack, but somehow there's never enough time to get it all done....

Herbert only appears to be soft. Last year, he successfully despatched king rat, but not before they had spent some time locked in mortal combat, with the rat hanging precariously from Herbert's snout.

I've known the farm for a long time. By a strange twist of fate my Uncle Alan (my mother's brother) used to work at harvest time for Joe's father. Today, the farm is perhaps a little crumbly, Joe's motto being 'put off today what you won't do tomorrow', but I'm in no position to judge. My den in the shed is just as chaotic. Surrounding me are various examples of woodworking incompetence. I start off with good intentions, with great care and skill, nails straight and true. The quality inevitably starts to slip, however, and the level of work-manship is embarrassing – especially when I compare my own endeavours to the bee's close attention to detail and structure.

My kit is piling up, ready to be transported to London, with only a small space for me to squeeze out of the doorway between the hives, honey boxes and frames, each awaiting a coat of bright bee-friendly paint.

Checking over and making up equipment

Carrying on from your preparations in January, this is the time of year to get out any of the equipment you've had stored since last season, and make sure it's all in good working condition. It's also the time to put together any new equipment which you've bought. The construction of beekeeping equipment does require some degree of space and some basic tools for assembly – hammers, a wooden mallet and an electric screwdriver being the key ones.

Like January, February is a great time of year to find some bargains as the larger manufacturers try and offer incentives to buy during the closed season. I think it is rewarding to assemble your own hives, and it allows you to become familiar with the various components and how they operate.

Peter, Zen master

Before the weather warms, my newly assembled equipment needs to be ferried down to London where it will form an integral part of my operation. First though, I do a recce, driving to North London to meet up with Peter Kingsey. He's an old bee-master, an ex-butcher and my mentor, who has guided me over the years. He is set to retire his old woodland bee site there and if I can convince him to let me take it on it will be an important stepping stone on my journey into the city.

It's a huge, sprawling site spanning three acres, with room for multiple hives, which Peter has used for queen rearing. Over the decade I have known him, he has granted me as much time and wisdom as I required, and I have gleaned, scavenged and scraped more information from this living legend than from anyone else.

From the first time I met him fifteen years ago, when I bought some bees from him and got to know him, I would regularly observe the old boy as he moved with quiet confidence among his many hives. He would softly sedate them with smoke from his pipe – a reassuring scent from a different era. He was steady and patient, his apiary organised and immaculate. He taught me that being gentle and calm was key; that I should avoid sudden movements, and keep my distance when I had a hangover or bad BO. In return, my hives would be happy and productive.

Smelling sweet Many people don't realise it, but bees have an impeccable sense of smell. Just as we do, they react badly to unpleasant odours; it makes them stressed, grumpy and more likely to sting their handler. To get on well with them you need to be clean and odour-free. Not always easy when you live a nomadic life.

An old hand Never using gloves, Peter would feel his way around the bees, ensuring none were squashed or inhibited. Thanks to the smoke from a gnarled pipe that always hung from his bottom lip, his bees appeared in a trance, uninterested in stinging him.

It is Peter I have to thank for allowing me to entertain my plans of an urban takeover. When he got in touch last year, I bought some bees from him and casually mentioned my interest in his North London plot of land. I knew he was ready to call it quits. He'd already made plans to retire to the Home Counties and was curious to know about oilseed rape, as he and his bees have never experienced it before and he is anxious about them foraging on it. Neither he nor the bees are going to know what's hit them.

Peter will always be a wily urban warrior, but beekeeping is physical work and when we meet this February he tells me he's ready to hang up his veil. He gets in touch with me not long afterwards, using Facebook (keen to show that you're never too old for social networking) to tell me that the site is mine if I want it. Peter is a shrewd businessman, aware that three acres of woodland in Greater London is worth a fortune, so there is some negotiation before we agree on the rent I am to pay him – but in fact it is a very generous offer and I can tell that he wholeheartedly wants me to succeed with my venture. The plan when the weather warms is to ferry bees into new satellite sites in central London using Peter's wood as a base; the Shropshire bees will all be delivered here first.

Renovating the old site To help me sort out the land, I quickly mobilise an army of volunteers for the coming Sunday, most of whom are city bankers and lawyers, and whom I have had waiting in the wings from previous urban beekeeping courses. It is a wet day when we meet there to begin bashing the site into shape. It is no longer how I remembered it when I used to visit Peter here in its heyday. He hasn't had bees here for a couple of years, and over time he has let the site become completely overgrown and neglected, with little light venturing between the ivy-clad trees – bees would suffer with dysentery and other gut-related illnesses if installed in this damp environment, for they love the warmth of sun on their porches and patios. Don't we all?

With its roof alive and growing, fuelled by falling compost from the giant redwood overshadowing it, the bee shed is also looking shabby. One side is open with over twenty windows decorated with colourful but now peeling window frames. Here, Peter bred the most remarkable bees – although he did produce a shedload of honey every year almost as an aside, the site was never really intended for honey production, just the raising of queens. He thinks that restored to the right conditions it could yield well over a ton of honey.

Over the course of the day, the land is transformed by my gang of eager city slickers, who relish the chance to get mud under their manicured nails. You can easily be lulled into thinking you are in the countryside here, amongst the dense vegetation, but every so often you hear a police siren that brings you back to reality.

Keep your hives off the ground

The following day – on my own this time – I lay out stands on which to balance the fifty hives that I'm going to bring here.

It's always a good idea to raise your hives off the ground. This is not just about allowing the air and sunlight to get to them, which makes for healthier hives, it's also about preventing the beekeeper from getting a bad back. I use breezeblocks on levelled ground, onto which will sit 8-foot pressure-treated wooden rails. I have experimented with numerous hive stands for bees, but have found these crudely assembled terraces work best and keep the bees at a manageable height.

When starting out, most people will probably use bricks or perhaps wooden pallets – the material's not too important, so just use whatever you have to hand. David has old metal-wired milk crates that work equally well, and the grass grows through them making a popular refuge for frogs and even newts that adore the protection. In Barnes, West London, a site where I had bees commercially some years ago, I would often find grass snakes under hives, sometimes several together. This would make me jump, although they were harmless. They just enjoyed the warmth the bees would radiate through the floor.

A new generation
The woodland haven is sprouting and shooting everywhere; bluebells are close to showing their colour and when I bring my son Ned to the site to finish off the stands, I can smell fox in the air. At seven years old, he is still nervous about bees and I am hesitant about forcing my craft onto him. At the moment, he enjoys pottering around this safe wood just as much. It is a sanctuary for all and I think the bees will flourish here.

We have a rough picnic in the back of my truck and fire up my stove and whistling kettle, which he thinks is very exciting. I adore the fact that my son already loves a good mug of milky tea like his dad, and a chocolate biscuit to go with it – life is simple really.

I'm often intrigued by what Ned tells his school friends about his dad's work. I asked him about it last week. 'You look after bees for people on their roofs, Daddy,' he replied. 'And you make yummy honey.' That's my boy. You can't fault his accuracy.

He also said that one day he wants to learn about the bees. Like a father who has spent his life as an actor in the theatre,

I've never forced this craft onto him but my God it's fantastic to hear the request. After all, he has only seen the bees a few times, usually when he was tiny and safely strapped into his seat in a sealed car.

In a bid to earn some cred with him and his school buddies, I've agreed to appear in a CBBC kids' TV show about bees and honey. It's being shot on the roof of a new bee site I am constructing on the South Bank, with superb views across the river.

So for now I'm hoping that some bee suits and empty hives should excite a few young minds. As yet there are not many young beekeepers but this I think will change as more schools have hives and integrate beekeeping into the curriculum. That has to be good news.

I've tracked down a small old veil and set of farming overalls with their pockets sewn up to ensure Ned is well protected. I even have some of my grandmother's old blue leather driving clothes – she had tiny hands – that I can gaffer tape onto his wrists. Poor lad. The outfit's low on style, but absolutely impenetrable which is what really matters.

I've seen old and young working together with bees in a wide range of places across the world. In Rio I once visited a beekeeper in the favela or shanty town – a retired naval captain named Gerard who tended his bees with a loaded rusty pistol wedged down the back of his trousers for protection against gangs. His young daughter, an essential part of the operation, greeted us at the door with a severely swollen face after an unfortunate bee attack. Her father simply dismissed it as her chance to build up her immunity against stings. I guess life's hard in the slums but I wasn't impressed.

Getting children started

If you want to get children interested, I'd suggest you look at the British Beekeeping Association's school packs and online information. It's great. I have involved my nephews with my bees since they were young and I'm hoping Ned will also take up the baton. I see that one beehive manufacturer now produces a virtual hive for teaching purposes, with photographs inserted into comb frames which show how the colony develops across the season. It's awesome – great for teaching beginners of all ages.

A ray of sunshine My bees are far more motivating than February's landscapes. London's parks look drab, muddy and lifeless when I come back to the city in the middle of the month. It's bleak, unpromising, hard to imagine under a wash of warm sun. But there is one ray of hope. Or more accurately, one ray of colour. It's a solitary purple crocus. Given time it will be followed by others, igniting this patch of sludge, and bees will start bringing in crocus pollen, helping to trigger good brood rearing.

It makes me think of Ann's sandstone walled garden in Shropshire. I'm sure its tiny warm sheltered corner will also be awash with early foraging bees, grabbing what they can in the narrow corridor of opportunity. Even the tiniest clump of snowdrops can become a mecca for foragers keen to gather some early pollen.

Not only does the crocus add a tiny speck of vibrancy as I walk through the small park. It also, I admit, instils in me a momentary panic. In common with many beekeepers at this time of year, I seem to have crammed a huge amount of work in already, getting ready for the start of the season. This lone bloom has made me realise that it is now almost upon us and yet I still have so much to do.

I'm back in London on my way to a business meeting. Potters Field on the Thames, which I always walk along to catch the tube, has seen substantial development in the past ten years and its light and openness inspire me. It's a rare chance to see the sky properly in London.

Its rejuvenation and rebirth into More London and the mayor's office makes it a popular spot for picnics, lunch breaks and photo opportunities. I know the management staff are enthusiastic to discuss placing bees on the site but it's not straightforward: not only is it heavily populated during the day meaning that the hives would have to be fairly high, its tall buildings and walkways also channel the wind into gusts, making it potentially perilous for bees returning from foraging.

At the same time there are things for bees to be thankful for here. Scrubby buddleias, crippled silver birches and scorched clover tufts have all been replaced by vastly improved, neatly coiffeured and permanently tended borders. They now provide an all-season, à la carte menu for the bees.

There's also a side serving of heavily scented flowering thymes, striking acacia flowers and numerous lavenders. They might be a less native range but they do provide a great mix of pollen over the season. It's encouraging news for the more discerning bee pallet, but I somehow doubt this level of posh Heston Blumenthal nectar is reaching every London borough.

I've heard that some bees are starting to produce small quantities of acacia honey in the capital. I'd love to send some samples off for analysis, but it's a very costly investigation, and as discovering their choice of nectar is based on pollen scrutiny within the honey, it's also not 100 per cent accurate: pollen isn't automatically collected from the same sources as nectar.

London's rapid redevelopment of these and similar areas also means I now struggle to find fresh wild green sites for bees. So I spend the last few days of February scanning more rooftop locations for my hives. I see this resource as potentially limitless, even if it means more sweat and tears trawling bees and boxes onto roofs.

What potential new beekeepers should be doing:

- New equipment should be purchased and assembled ready for the start of the season.
- If you are serious about keeping bees, you will need to order a set for the summer as breeders close their books early.

More extensive beekeeping tips:

- All your equipment – both new and refurbished – should be almost complete and ready for the onslaught of the season.
- Keep patience and don't disturb your lovelies just yet. Do, however, check hive weights and make sure entrances are not blocked with bees that have died naturally.
- On the first warm brighter days, look out for bees foraging on early pollen from snowdrops and crocuses. This will signal that the queen is in lay.

MARCH

COMING BACK TO LIFE

For both honeybees and beekeepers, March is the month when we begin to stir from dormancy. Over the winter months, it's been hibernation, or perhaps more a catatonic slumber. For me, it's meant an extra-togged duvet, while the bees have been kept snug with recycled cavity-wall insulating boards on top of their hives – scavenged from skips on building sites and stuck to plywood with gaffer tape to prevent the bees from nibbling the foam away. This is another classic example of beekeeping make-do-and-recycle and I think the bees appreciate the extra homely touch.

For much of the winter, apart from the odd foray to London to get ready for the bees' and my arrival, I have been holed up in my temperate bee shed in Shropshire, listening to the cricket in India on *Test Match Special*, drinking tea with my dear friends in Joe and Ann's parlour and building new hives. It sounds cosy but it's incredibly hard work building hives, as anyone who's ever done it will know. Low points have included nailing my finger to a workbench with a nail gun, not just once, but twice, as I battled with the fiddly components.

The problem is, I become dazed by the repetitive business of assembling honey frames – something you too will grow to hate. I've managed 3,000 bloodstained frames for the season, enough for 300 boxes. There's nothing like being optimistic about a gigantic honey flow. All those years ago when I first installed my hive in Bermondsey, the honey just kept on flowing in all summer, and every year I hope for the same. In any case, it's always good to have extra equipment on hand – much less stressful than having to make it up in the height of the season.

Beekeeping is a passion as much as a business and one of its compromises is a nomadic lifestyle. For the last few years, in order to obtain unique, subtle flavours in my honey, I have been moving hives around the countryside, in hot pursuit of flowering dandelion and bluebell, blackthorn and lime nectar, while also doing my best to avoid mono-floral crops such as oilseed rape. I think my customers appreciate it and there has been a real move by them away from honeys whose labels say they are the product of more than one EU country or even non-EU countries. It's heartening to think that this might one day be the future for all honey.

I'm encouraged to hear customers wanting to know about the traceability of products and that goods have been produced fairly and with consideration. After years of painstakingly separating my honeys into regions, boroughs and seasons, I've been rewarded by a devoted following. This March there is a rise in demand for freshly drawn spring honeycombs, white and light, as a refreshing tonic to lift the body and mind – the order book is swelling already. This is a product, however, that will only come to fruition with prolonged fine weather and, of course, bees on my new urban sites.

With the weather warming daily, the wildlife in Peter's wood is beginning to stir.

The foxes have been active – I can smell them when I arrive, their scent thick in the still air. Freshly dug earth has been scattered around the area in which the hives are to go, such has been the velocity of their digging on the overlooking bank. I'm not worried, though – they won't pose any threat to the bees.

I get ready for their imminent arrival here in the wood by mixing up a huge amount of sugar syrup just in case the bees are hungry. Forewarned is forearmed: this is fuel that may be needed by my lighter colonies when activity starts to increase.

Some people believe that feeding syrup to bees is hypo-critical given that for most of the year, you are selfishly steal-ing their honey for yourself. This is a fair point but it is also common in our terrible climate for some colonies not to be able to produce enough honey stores to last them through the winter. In that case, not feeding them would result in bees dying, and that seems to me a much worse crime. If I feed my bees, I do it knowing that it is essential for their existence.

I'm hoping that these North London bees won't need feed-ing, but I have to be ready in case they do. Things can change very rapidly at this time of year and though I wouldn't normally put syrup into the hives until April, I don't want suddenly to find that the bees are starving and I have nothing to give them. This syrup never makes it into the honey store; it is simply designed for brood rearing and supporting the colony.

Making sugar syrup

To make my sweet syrupy brew I raise an old open-topped drum up on bricks, pour in a measured amount of rainwater that I've collected and filtered from the shed roof, then heat

47

it with a large builder's gas burner. It's not ideal but as yet I've no electricity or running water up on the hill – and in any case bees like rainwater best. I'm investigating the possibility of fixing up solar panels for lighting the bee shed instead of my useless wind-up head torch, but that's as far as it goes. There are greater priorities at the moment. And part of the fun of beekeeping is making do and being inventive.

If you want to make your own syrup (you can buy it ready-made but it's very expensive), I'd like to raise a red warning flag right at the start. I can't stress how crucial it is to keep everything to do with honey – and I mean absolutely everything – away from your syrup mix to avoid any possible contamination. Even the barrel or pot you make it in must never have been used for honey storage, as brood diseases can remain inert for years in the form of spores.

My measuring technique is rudimentary. I use a nail to scratch a crude mark inside the drum, indicating the correct amount of water. I always wait for the steam to start rising – you only need to warm the water enough to dissolve the sugar.

Now for the fun bit. In the back of my truck I have ten 25-kilogram bags of British granulated sugar from Brick Lane's Bangla City hypermarket – thanks to some honey trading I've negotiated a great price on it, and it's going to give me around 400 litres of syrup. I'll make this in two batches, but if you're making smaller quantities of syrup in your kitchen, use a solid pan and buy 1-kilogram bags of sugar at the best price you can find.

I always manage to slosh syrup everywhere, so at this stage I don a giant plastic apron made by the late 'Tarpaulin Mike', whose skilful hands over the years have made trailer covers, bee-proof nets and much more. [Sadly, Mike died last year. My apron is now part of the legacy of a great craftsman.] This apron was made in minutes and stretches to my toes, with a length of bailer twine around the waist – it's a piece of kit that is full of memories and practical at the same time.

Add the sugar a bag at a time to the warmed water, stirring each one in fully before adding the next. The consistency of the syrup's dilution depends on the time of year. I use a thinner brew for spring feeding and a thicker mix – closer to the consistency of honey – for young nuclei.

It's vital that the sugar dissolves completely in the water, or you'll end up with crystals that stick in the feeder later – you can check for crystals on whatever stick or paddle you're using to stir in the sugar. I use a giant wooden paddle also from the Bangla City store, which is perfect for keeping the brew moving. You can tell you are getting close to the correct consistency when sugar lumps start to float on the surface and are harder to dissolve. I like to run my hand through the mix to feel the consistency, but do take care first that it is still cool enough.

Bear in mind that you're not making toffee. Don't boil the water, and do take your time. If you're in too much of a hurry the mixture will overheat before the sugar has completely dissolved; if you boil it after the sugar has dissolved, you'll get toffee.

Once the fresh mix has cooled I store it in 25-litre barrels for portability, seal it tightly and label it with the date and the type of mix. Now it's ready for spring feeding.

Back with my little darlings   It feels as though spring is only a few weeks away – and that will mean the chance to get my bare hands back amongst my downy bees. Small wonder I'm energised! I can hardly wait to get stuck in again. No doubt there'll be a brand new set of adventures leaving me emotionally drained. It's not good for my romantic life but what the heck. With this level of devotion and conviction it's virtually impossible for me to have any ladies other than queens in my life.

Back in Shropshire, I resist the temptation to delve straight into my hives. It's best to leave well alone until the warmer weather has officially arrived. Only when temperatures outside reach a constant 14 degrees – usually towards the end of the month, about the same time as the clocks go forward – should the beekeeper consider opening up hives or moving them around.

Peter advises those who want to mother their bees to go and buy a dog or cat. Bees can be damaged by constant attention. At the very least, visits can disturb the nest's delicate balance – and there are those who believe inspecting a honeybee's nest is 'raping its privacy and independence'. This I feel is a little extreme. Some of my clients would not be happy with bees swarming everywhere, which could happen if I failed to inspect them in summertime. It is clearly a very delicate balance.

You don't want to chill or cool a hive by opening it up in March when the weather is far from consistent. You'll only knock the bees back in terms of development, as it takes a while for a hive to recover its natural harmony once it's been opened to the elements.

Of course you can sometimes see what's going on from the outside. Bits of wax and various body parts from stores being uncapped lying underneath your hive, for instance, can be a sign of the first movement. An increased activity at the front of your hive, with huge quantities of pollen being brought in, will signal not only an explosion in population but also the fact that it's time to remove the winter entrance blocks as the front becomes congested with traffic.

I find myself waiting for the warmer period with huge excitement, monitoring weather websites and local forecasts. Warmth is the key to a good beekeeping season, but as well as this you need moisture. It's a fine balance and one you can't control.

For most of the winter the bees will have been clustered tightly in a ball, awaiting the longer days, in a hive maintained at a constant temperature of 32 degrees. The bees all shiver to keep the temperature stable; there are also special heater bees, whose role it is to vibrate inside the combs, radiating heat for the young brood. Without them these could not otherwise survive the harsh winters that have become more frequent in recent years. Very tiny patches of brood – sometimes just the size of a fifty-pence piece – can exist throughout the winter and help to regenerate any losses and maintain numbers. In extreme weather brood rearing will cease completely as the bees remain in a tight cluster.

This is a time of great regeneration for bees. Those who have taken care of the queen over the winter months will die off and a new generation emerge. Nursing the queen and tending to her every desire over the winter period, these dying bees have nevertheless worked less hard over the winter, become less ragged and therefore survive over the six months until the new brood can take over the devotion. It's a big changeover. The new spring bees will have a significantly shorter lifecycle, lasting approximately thirty-six days from their emergence, as their colossal workload intensifies.

Like the bees, I have also had to work hard to keep myself warm. In my bee shed I have relied on a stove from the Falklands with an enamelled grate, bought on eBay, which has rarely been without a giant kettle steaming away on top of it. I've been burning timber pruned from various apiary sites; it would otherwise be shading the bees in these darker months from precious sunshine, which is essential for disease prevention. The wood is full of sap and a little too green for burning; it spits and hisses as if alive.

Although I have missed my bees during this bleak separation, their sound and their smell – as you will too if you start beekeeping – I'm under no illusion: this is a one-sided obsession. When I do eventually crack open the hives for the first time – releasing a strong musky odour as the bees bubble out from the frames, tumbling on top of each other to see what has awoken them from their drowsy state – I know they don't feel the same way about me. Later in the season, I begin to sense that they recognise me. 'Here comes the beekeeper,' they probably think, but at this point in the year, I'm nothing more than a stranger or even an intruder.

What you need to wear

Early in the season, bees tend to be gentle and easy to handle without gloves or smoke, but you should still always wear a veil as stings to the face are most unpleasant. The first few times I have contact with them, I tend to be too relaxed and the result is several stings to my fingers. Later in the season, immunity will kick in, but for now my digits swell into ruddy chipolatas; we are not yet familiar with each other and acquaintances need to be fully renewed.

If you are starting out you will no doubt, like me all those years ago, want to wear full battledress whilst you are entertaining your bees. Your confidence will definitely increase and over time items of clothing donned will be reduced, enabling you to do a better job. If you're not wearing massive gauntlets, for instance, it's easier to feel when bees are being squashed. The result? A calmer hive, fewer stings, and a more relaxed me.

In the winter, for a quick check of my bees, I might use a smock veil of the type that pulls tight around your waist. Veils and hoods can vary considerably and I favour a hood in a fencing-mask style – mine now comes complete with a piece of gaffer tape over my right eye, where I burnt a hole once lighting my smoker. I would suggest adding a cap, if you have long hair. You will also need some good gloves. I prefer leather ones in the summer – calfskin by choice – because they breathe, and light rubber gloves in the winter.

If you want to be less of a summer's day beekeeper and perhaps help others with more bees, you need to be in total control and have suitable equipment, in which case I would advise you to get some white overalls and a veil that fastens underneath the suit. This is what we use commercially and it is the best system for a day in the field.

Always try on new overalls and veils to check they fit – but not too snugly – and cover you fully. Luke, David's assistant, has a set which makes him look like MC Hammer, but baggy is good as the bees are ingenious at finding points where the suit touches your skin and they can sting you through it. I know that a former pupil of mine, Esther, wore her new outfit to a fancy-dress party when she was starting out to check it was comfy – she looked great on the dance floor, whirling around.

Here is a top tip for when you get a bee in your veil – no ifs about it, this will happen. Some of you will be alarmed to hear this and perhaps I shouldn't mention it, but it is important to know what to do when the inevitable moment arrives. It happens when you haven't quite zipped your veil fully or when you've developed a hole, but more likely it will be when an ingenious critter has crawled up your trousers and suddenly appears in your eye line.

It sounds obvious but the first thing is not to panic. Sadly this bee has to die because opening your veil when you are working your hives is not an option. I generally find intruders like this will fly towards the light and the front of the veil. With your gloves on you just need to squeeze the bee, pinching it in the veil. In most cases, I find the bee has already shed its sting somewhere on your clothing, so it will ultimately die in any case – but having a bee up your nose or a sting on your eyelid is not pleasant.

Varroa and disease

It is comforting to have intimacy with my bees again, but it's a bitter-sweet feeling as there is winter's damage to assess and possible outbreaks of disease to be dealt with, which sicken my heart. Disease often rears its ugly head at this time of year. Where whole colonies are lost, hives should be removed to prevent their takeover by spring inhabitants looking for nests to raise a family, and sterilised along with the frames ready for reuse.

As we have seen, Varroa is the big killer. Like HIV in humans, it is not actually the Varroa itself that kills the bees but the other viruses that they can catch when they are weakened by it. There are any number of diseases that can affect bees and writing about them would make a whole separate book, but a horrific one is Cloudy Wing Virus or Deformed Wing Virus, which Varroa-affected bees are very prone to, and which prevents a bee's wings from developing normally. Emerging bees with stunted and malformed wings will be unable to forage properly for nectar – sometimes they can't even leave the nest – and in time the colony will crumble, often as a result of starvation. There are loads of books and online information about this terrible epidemic.

Monitoring is key for a beekeeper, to keep you abreast of any damage or infestations. It is worth saying again that the system that works for me is using oxalic acid treatment in January (see page 17) – though I'm sure it will change as the mite continues to evolve and new, more resistant strains emerge. But the mite numbers will be greater in the late summer, so a good time for a proper test – the icing-sugar shake as devised by the experts – will be essential at that stage. Details of how to carry out this test are given in my chapter on August (see page 165).

Website for bee disease To help you keep informed of all the new invaders, have a look at the website of the Central Science Laboratory, currently part of the Food and Environment Agency; they have a National Bee Unit and they give regular disease updates and lots of advice.

Local associations also run disease awareness courses, which are fascinating – honest and worrying – but well worth attending.

Basic hygiene

Hygiene is crucial in beekeeping. In order to prevent cross-contamination of brood diseases, bee inspectors today carry washing-soda buckets with a splash of bleach to clean their hive tools between inspections of individual hives. It is vital that beekeepers themselves work as members of a community, registering bees and preventing disease, especially in urban areas where bees are kept in close proximity to each other. Their concentration appears to be forever on the increase – it is a distant cry from those early days in Bermondsey when the only bees I would see around were my own banded beauties. Though I applaud the fact that so many people are now keeping bees, it does mean we all need to be that bit more vigilant.

Another rule to remember for keeping disease away is never to feed your lovelies honey from other bees. They should, of course, have their own stores, or you can make up syrup for them (see page 47). I heard of a vicar in the Midlands a few years ago who decided he would give his bees a sticky treat and bought some South American honey from a supermarket. By the time his colonies were inspected in the summer every one had been infected with AFB (American Foul Brood) – a catastrophic spore-based disease that can remain dormant for years within hives and honey alike.

Each one of his hives had to be completely burnt to prevent the disease from being spread and then buried under several feet of earth – to stop any other bees getting hold of the honey that might have wept from the burnt combs. There is no other option with AFB. European Foul Breed is slightly less bad and you might possibly be able to cure an outbreak with antibiotics, or by shaking the bees onto fresh frames – called the 'shook swarm' method. The important thing to remember, though, is that if you suspect you have either AFB or EFB you must report it to the National Bee Unit who will send out an inspector. This is vital for everyone's sakes.

I'm worried my own bee fatalities will be greater this season due to the new way I have been treating them for Varroa. For now, I can see that several hives have already suffered and the colonies are lifeless; the damage I fear may have started last autumn.

Don't look up On a warm day, I should also warn you, March and early April can be one big poo fest. While it is an incredible thing to see the first major flight of the year, you don't want to be underneath your bees as they release several months of crap.

These hygienic creatures hang on to their number twos until fine weather allows them out. When I was a kid, they always seemed to target my mum's white washing with their streaks of toxic yellow excrement. Or they waited until just after my father had washed the car. In the eighties he was convinced his Austin Allegro was the victim of acid rain. Sorry, Dad, it was bee crap that you had to chip from the paintwork.

On the plus side, the warming spring days can be good for the soul, and I try to take a minute occasionally just to soak up a few rejuvenating rays. It probably wouldn't do me any harm to have a sip of my friend Lara's grandpa's delicious restorative at the same time.

Dr Bernays' Honey Gar

I remember my grandfather, who was a doctor in Chichester, sitting in the afternoon sun with a tumbler of the honey gar and carbonated water which he swore by as a cure for all ills. He recommended it to his patients as a refreshing tonic to lift the body and mind. A glassful to be drunk once a day for good health and longevity – doctor's orders! With all the amazing things we now know about honey and cider vinegar I put this down as a wise old tonic.

Runny honey
Cider vinegar

Empty one jar of runny honey into a jug. Fill the jar with organic cider vinegar, swirl it around to get all the remaining honey, and empty into the jug. Mix and pour into a glass bottle. Shake well, then store in the fridge. When required dilute with water like a squash. Carbonated water is what Grandpa used.

For an instant honey gar simply place equal quantities of honey and vinegar into a cup and mix well with water.

On the road again   If the season is advancing in Shropshire, I'm sure it will be galloping ahead in London where I hear it has become milder. I use the bee webcam on Fortnum's website to gauge the activity of the bees – it gives me a rough picture of the numbers heading out on foraging expeditions or taking away debris on cleaning relays. Although I am a real traditionalist, I feel it's important to embrace the new wave of available technology; it makes my life less hectic and provides me with some fascinating facts and figures.

I'm keen to get on with my big bee haul. Moving bees is easier when the weather is still cool and you don't have to worry about hives overheating; you're also less likely to leave stragglers behind if they're mostly still dormant.

It's time to begin the big move south....

Having packed up my various bits of bee kit on the trailer which I've borrowed for the journey from my mate Henry the blacksmith, I set about loading the hives onto the stainless-steel checkerboard flatbed of my bee truck, a converted Toyota Hilux. As usual I start by consuming some bitter dark chocolate for instant sugary energy. Then I begin with the heavier hives, which are furthest away within the apiary, on the bottom layer, after which I'll load the lighter ones on top.

This means that as I become more exhausted, which I inevitably will, the journey from hive to truck gets shorter and therefore both physically and psychologically easier.

I keep the roof of each thrust tightly into my chest, whilst my hands grasp around the floor. This carrying technique is crude, but it's a method I have mastered over time to ensure that the roof remains firmly in place – thus guarding against a mouthful of bees.

Though shifting them is still a back-breaking task, the hives generally feel on the light side. This signals that the bees' stores of honey are significantly low, and they will need feeding if the warm weather doesn't arrive soon. It's a good job I have some feed ready at the wood. If their numbers are to expand, the bees must have enough sustenance – but that will have to wait until we reach London.

I notice one of the hives is marked on its identification plate as having one of Peter's own queens in it. So it'll be a homecoming for this chubby lovely, as she would have mated high above the treetops in North London with numerous drones a couple of summers ago. This could be her last year of production as she begins her third season.

Usually this third year will be the final one for my queens before they become infertile and are replaced by the bees themselves, or by the beekeeper, in order to keep the colony virile. But perhaps her legacy will continue; if her character-istics are impressive, she has a good temperament and her offspring gentle and hard-working, she could be viable stock from which to graft eggs later in the season, in order to create magnificent daughters. These are some of the attributes I consider when selecting queens to use as breeders.

Working on my own, I take time to load fifty hives onto my bee truck. Even though I'm travelling in the dark to make the journey as cool as possible for the bees, the hives need to be very carefully stacked to allow maximum ventilation through the vents in their floors and around the sides. The truck has been specially designed with this in mind.

I'll be hitching Henry's trailer on to the back, and strapped to the headboard behind the cab is my ageing 1966 ex-Italian army Vespa – now looking a little weary, but ready for a new lease of life. Traffic has become so bad in London and I need to look at viable ways of getting around town easily to view my bees in rush hour. I'm also thinking perhaps I could get a sidecar. This would be handy for carrying hives around in and it could also be great for market stalls, with a simple umbrella and table.

Strapping down hives is a skill, and it can take time to perfect in the dark (perhaps on a moor), when you are already exhausted from loading up by hand. I always pull over to check the load under the first light I can find. While I'm doing this I have to be careful of disorientated bees that might be clinging to the fronts of hives or underneath them – obviously I'm illuminated too and they can spring an unwelcome attack.

Touch wood, I have only ever had one hive fall from a badly tied-down load. It was on a steep incline and the load had slipped – I heard the thud at 2 a.m. and jumped out to find the hive upside down. I was on the Shropshire hills and I was fortunate that time because the roof had stayed on – although the bees were very grumpy afterwards, the hive went on to produce a good crop of honey.

I think every beekeeper, on any scale, has a disaster story to tell about an incident with their bees … but when it's being done commercially, there's an added intensity of pressure. We sometimes find ourselves on a knife's edge with energy and emotion, which can force a bad or ill-advised decision. Throughout the season we always seem to be rushing about on winding B roads, and more than once I have flaunted with meltdown in the early hours.

Occasionally I have inadvertently dropped beehives in the middle of the road when running across from a gateway to load them on to my truck. I have also once reversed the bee truck over hives on a heather moor in the dark, thinking the bumps were being caused by my wheels going into rabbit holes. David took that one very well.

I have even briefly nodded off at the wheel when driving to London – this was somewhat terrifying and I now ensure I have slept well before trips and am loaded with caffeine. Apart from any other unfortunate road users, a totalled truck of bees would be an unrivalled catastrophe.

I have considered a more sedate way of transporting hives around London, with the use of an old milk float … but charging the ageing beast would be a problem. I plan initially to use the truck's flatbed to ferry them down from the North London site to their various central city locations as it can move large numbers of hives relatively quickly.

Everything on my bee truck is bolted down and welded with solid steel. The result is a vehicle that looks armour plated, almost military in style – it would look at home in North Africa with a giant machine gun mounted on the rear – and I adore it. Holes in the front bodywork of the removable canopy allow air to be channelled inside to help cool bees on their journey, essential on humid nights. And a trunk is attached above the roof of the cab for foldaway chairs, extra supplies and water barrels – for washing and of course tea making.

David has started taking holidays – it's something he rarely did before, so this is brilliant news – and he came back from a trip to Australia recently with photographs of various utes. A ute is basically a utility vehicle with a flat tray on the back but in Australia it's taken a stage further and a canopy is also added, which can be zipped open or closed, all for experiencing the great outdoors. I modelled my truck on these photos and it was built by Henry.

On board this British ute this evening are all of my basic possessions – my vintage bee books and clothing, along with my honey extractor, honey boxes, a sleeping bag and tent. Sadly no cats – I leave my pals Hinge and Bracket sitting on the drive as I depart in my nomadic caravan. Their future is assured with a glorious rambling country estate as their playground and a limitless supply of rabbits. Mine is a little sketchier.

I have been offered a few floors to sleep on but accept the one from an actress I have just started seeing, although I'm conscious she might not be fully aware of the amount of bee clobber that comes with me. The North London shed will be handy for storing much of that. I keep with me a change of clothes and essentials in my father's old leather army suit-case, stencilled BENBOW, which he used for national service in Cyprus. It offers huge comfort, but when I arrive on anyone's doorsteps I do look as though I'm selling ency-clopaedias. I long to see it empty as that will mean I have settled somewhere.

I cautiously navigate the winding tree-lined driveway of the sprawling country estate, toot the horn at the cats I can see in my wing mirrors, signal right, then it's on to the main carriageway, next stop North London. But first where's that Best of Ibiza CD? It's time to play some driving music.

What potential new beekeepers should be doing:

- Begin positioning your new hives in situations you feel will be best for your bees. See if they catch the morning sun and are sheltered and away from neighbours and pets.
- Having learnt the theory, enrol on practical taster courses and get in touch with your local association – membership is usually due in the early spring and you need to find yourself a mentor.
- Try on new veils and overalls to check they fit – but not too snugly – and cover you fully.

More extensive beekeeping tips:

- Mix up sugar syrup in advance, in case you need any for spring feeding – as soon as the weather warms and bees become active they may well need a sudden boost.
- Double-check that all your equipment – both new and refurbished – is ready for the imminent onslaught of the season.
- Keep curbing your desire to get stuck into your bees. Watch the weather forecasts and don't disturb them until it's warmer. Use the time instead to observe any life around the mouth of the hive.
- Be prepared for disease and death in your hives at this time of year. Keep a close eye in particular on levels of Varroa infestation.

APRIL

BACK TO URBAN SPLENDOUR

It's 2 a.m. and I've just arrived at the North London site. I crash the truck through the undergrowth, accidently taking off its radio aerial in the process by swiping a low laurel branch, and at the same time buckling the bike belonging to my actress friend which is strapped to the roof canopy. So much for a stealth arrival. With the truck's spotlights blazing on the path and me soon parading around in my bee suit, the neighbours are sure to think aliens are landing.

Fortunately, there aren't too many onlookers – not at this time of night anyway. It's got a mixture of open space and municipal buildings all around so neighbours are thin on the ground. Although they're exhausting, middle-of-the-night missions such as this one are all part of nomadic beekeeping if you move your hives about or have them in more unusual locations around the UK.

This is because you can't start a journey until the last bees have returned to the hive – you don't want to leave behind your best workers – which only happens once the sun starts to set. So by the time you have arrived anywhere, you'll be fumbling around in the dark. I opt for a head torch

– a wind-up hippy one which is very PC but a bugger when the power drains just as you are carrying a hive through dense undergrowth. The last thing you want to do is put your foot in the wrong place in total darkness.

Before I can snuggle down in my sleeping bag in the back of the truck, I have to place all fifty hives in their new location. I heave the first one out, staggering under its weight. Moving bees at any time of the year is back-breaking stuff. It should be easier in April as the hives are lower in stores and therefore lighter, but my body is not yet conditioned for it and I strain with the first few boxes. But David always says this is the easy bit of the journey – he doesn't notice the weight because he is already thinking of his bed. Let's face it, he's moved thousands over the years.

The hives take over an hour to unload and arrange, even though I've already prepared the stands by placing them in the correct position when I was here a few weeks ago. For carting hives around on bumpy ground, particularly if the very last leg of a journey doesn't suit a vehicle, I use a modified wheelbarrow. It sounds unlikely, but it works best without the barrow part so it's just the wheel and the frame, on top of which I place a hive. It's safer than carrying them when negotiating tree roots and small holes dug by badgers. The last thing I want is to flounder around and upend the entire contents of a hive.

Finally, I collapse in the back of the truck for a bit of kip, but not before brushing the area down to remove any sleepy bees that might otherwise find their way into my sleeping bag. A bee in your bedding is a worry, and the sound of buzzing seems intensified when you are trying to sleep – you simply will not rest until the intruder has been evicted. I use cardboard underneath my bed roll for extra comfort and I always have a pillow hidden somewhere on board for that little bit of luxury. Like a dog, I curl up and settle for a few hours' sleep. Rolling onto my back I feel a burning sensation in my shoulder – I've located one of my lost bees. The pain of having her sting implanted into my skin is short-lived though, and I groan, pull my sleeping bag over my head and dream of a bee paradise.

By the end of the year, I'll be much stronger and tasks like this will feel less draining. I'm built like a whippet but beekeeping keeps me fit and my arms carry tons of equipment over the year – no need for a gym membership, every muscle gets a daily workout across the season. I call it my fighting weight.

Moving and positioning hives

When you move a hive, the golden rule is to make it either less than three feet or more than three miles. Bees are sensitive to the earth's gravitational pull. They are also able to pinpoint their home's location in relation to various landmarks; with their compound eyes they can see the sun through cloud in poor weather, and will use this as a positioning point. Moving their hive more than three feet but less than three miles will result in thousands of confused bees gathering around where their hive used to be. There they will stay and wait indefinitely until they die.

It's a tragic sight to see confused bees waiting where a hive used to be. It means if you want to move them from one end of the garden to the other, or from a rooftop to the garden, you will need to move them at least three miles away for a few days first, so that they forget their original location, which has been embedded in their psyche, before installing them in their new location.

67

Clearly, this is a tremendous faff so it makes more sense to position your hives extremely carefully first time round. Planning is crucial. To find the best place, I use a battered old compass to help me locate a south-facing plot that will provide lots of sunshine. It's important to think about the fact that the sun does not rise as high in winter and can sometimes not make it above trees or walls.

It's incredible how a little bit of sun can help the bees to thrive. As well as warming the hive, it aerates it by drying out any moisture within. This is a good thing in terms of preventing nasty bee diseases such as Nosema, which attacks the intestinal tracts of adult bees. Sun also warms the bees' flight muscles and gets them up and out early in the morning so they'll produce more honey for you. Just like humans, I think sunshine keeps them happy.

Moving bees in the winter or early spring is advisable. Not only will they have time to become established and therefore productive for the main honey flow which comes later in the year, but there is also little chance of them overheating on the journey. You still need to be cautious, however. Between November and February, bees could be clustering and in a delicate state. If you bash them by driving vigorously over a pothole, you can knock them off the frames and they'll all end up on the mesh floor of the hive where they'll block the small holes which allow ventilation.

In the summer, heat is your main problem, and it is more catastrophic. Bees can collapse and overheat in a flash, another reason for exhausting yourself by doing the move in the early hours while the temperature is cool. Although I always try my best to find a good position for them, sometimes when I'm really tired I have been known to simply dump a hive in the easiest place and keep my fingers crossed that it is not a damp shady spot. Usually it works out OK, but I kick myself when in daytime I see the perfect spot that I missed in the dark.

The majority of beehives today now have mesh floors on the bottom; as well as helping to filter out parasites which fall from the bees during treatment, they make moving the bees easier. In the past I used to use travel screens, which could be screwed on top of the nest box, or brood box as it is more commonly known, which would allow plenty of air across the bees in transit. The problem is that bees can get very excited when these are fitted and they offer little protection from the elements if you move hives on an open trailer or flatbed truck.

Whether you use travel screens or have mesh-floored hives, a piece of carefully shaped foam should be used to block up the hive entrances when you're fairly sure all the bees have returned home. I also carry dozens of rolls of gaffer tape, to plug up any gaps and seal in leaking bees. It is worth noting, too, that there always seem to be some congregated underneath the hive and for this reason I always wear gloves and gauntlets when moving them, having learnt the hard way – no matter how many stings you receive on the hands, the pain must be endured as you simply mustn't drop the hive.

My hives are secured before a move with brass stable pins which you can buy from beekeeping stores. I would recommend all beekeepers use these. Unless your hive is of a design which means the mesh floor is already attached to the brood box, for instance, you will need some kind of strap to keep it all together.

And unless the hive's roof is firmly secured to its body, the complete unit can slip, depositing 50,000 bees in your face. With this in mind, it is always best to wear a bee suit during any such manoeuvres.

Some people just can't master the technique of carrying a hive and you may find you have to rely on someone else to move your hives with you. Look for long arms when choosing your assistant.

Constantly bending over hives is big trouble for my ailing back (my osteopath Brent was keen to come on one of my courses until he gave me a physical assessment). But for most beekeepers, lifting and moving hives is not a regular duty.

Feeding your bees

One time when you will need to lift a hive is to gauge its weight, and by doing so work out whether your bees have enough stores to keep them going. This is called hefting. It can be hard to assess but you'll find it comes with experience. The best equation is that you'll need to put some effort into the lift if a hive is heavy with stores. Doing this means straddling a hive and then gently lifting it up between your legs, without any shaking. In large quantities, bees weigh a surprising amount. If a hive can be lifted too easily, there is a chance its occupants are extremely hungry – which is why at this time of year, it's so important to inspect them regularly. And if they are hungry, whatever the purists may say, I think it's time to feed them. Better that than let my lovelies die of starvation.

If the hives are light, they may be easier to move but you will then need to lug around huge heavy flagons of sugar syrup. If you've been organised, the business of making it should already have been done in March. But if you need some in a hurry you can do it quickly by mixing British granulated sugar only with warm water, and keep it in one-litre plastic milk containers.

If you are transporting large amounts of sugar syrup, as I do, make sure the lids of the containers are on well. I even use gaffer tape to seal them fully, as a bump in the road can lead to a very sticky truck. This is fine out in the middle of the countryside but not in an urban area – on a warm spring day, you will instantly have a truck adorned with crazed bees, ready to take full advantage of your mishap.

Sometimes, as a beekeeper, you find yourself coming up with some odd solutions to extreme situations and several years ago, I knocked up a hoist and pulley in order to move a colossal tank off my truck. For doling out larger sloshes of the syrup, I avoid completely full buckets; not only does this save a tired back, but an only half-full bucket gives me greater control when I'm lifting it over the hive. While like me you will probably find yourself and your overalls getting sticky or crusted with dried sugar, be careful not to splash the syrup when you pour it into your hive. Any spills will attract other bees which could rob your colony of its honey.

To give the syrup to the bees, I use a Miller feeder in each hive which sits directly on top of the brood box and frames. It has two chambers so that the bees can climb up into either of them from a gap in the middle. The syrup does not leak because each chamber is waterproof, having been dipped in paraffin wax. Each chamber also contains mini rafts for the bees to clamber onto, in the form of stringy wood wool, as otherwise they could easily drown. The rafts help them paddle around and lap up the sugary liquid quickly.

It's worth investing in a proper feeder, because you don't want to cause a sticky mess in the hive. Once you've installed it, make sure there are no leaks around the edges. Oozing syrup can send the bees into a frenzy as they all get very excited and squabble over it.

When it's chilly, bees are not able to handle sugar syrup. Not only are they not keen to go all the way up into the feeders in the cold, but the syrup itself can crystallise and freeze in their tummies, killing them, or sometimes it gives them dysentery. Instead, during the colder months I feed them a baker's fondant, the old-fashioned icing that you find on an iced bread bun. Fondant can be placed directly on top of the frames and the bees can crawl inside the plastic bags to feed. If you strike a deal with a friendly baker, the fondant is pretty cheap and it is also soft and easier to handle than syrup.

Once the weather has warmed, however, and the bees are flying around in the spring, syrup, closer in consistency to nectar and more easily handled by the bees, is the best way of getting stores into them quickly.

How have my bees fared? I should warn anyone embarking on a first bee adventure to expect losses. Natural wastage over the winter months can take its toll on your hives, and you shouldn't let this dishearten you. In addition, disease is on the increase due to overuse of chemicals and the rise of badly managed bees. Bees, like people, live in close proximity on this tiny island. If colonies are not properly maintained and pick up infections, then when they are visited by clean bees the disease spreads easily.

Counter-intuitively, a harsh winter does not always mean a decline in numbers. It can mean your bees emerge with less disease as harsh weather is thought to kill it off. And if some bees do die off then there's less chance of the whole colony dying of starvation, because there are more stores to go around the survivors.

Each year, I pray my losses are not high in number. Building up existing stocks or creating new colonies from them will mean a smaller yield because honey production has to stop for brood and bee rearing. When you remove a few frames with broods to start up new colonies, this temporarily weakens the bees because you are splitting them up. In the long term, you will have more hives; but in the short term, there will be less honey. Even buying new colonies will mean waiting another year before they are ready for full honey production as it takes them a year to gain full colony status.

Over-wintered queens

If you have ordered new bees for the season, and especially if you're a beginner beekeeper getting a whole new colony, then they probably won't be ready for collection until June. Now, however, is the time when an experienced beekeeper might be getting some over-wintered queens. They're better developed and more likely to give honey this season than new young queens. If you know the breeder, you can generally just ring up and find out what's available.

I think it's always fun to go and meet the breeders – that's how I got to know Peter in the beginning – but sending bees in the post is also possible. Obviously the boxes need to be well ventilated and the packages clearly labelled so that they are given adequate space and air during transportation; it is vital not to overheat them. A young nucleus of bees can be sent in late spring; equally queens can be delivered by post in cages. This is nothing new and it is on the increase again with the greater demand for honeybees.

Some of my queens come from a marvellous bee breeder in Pembrokeshire, who is looking at sending queen cells in the post this year. These sensitive cells need to be handled with particular care as any jolt or knock could damage the queen's development, especially her wings in the last few days – a virgin queen with badly developed wings will be unable to leave the hive to mate and will be killed off by the other bees. Nick thinks that if he sends them with a few frames of bees and uses a good courier company they should arrive safely. It would save me the round trip from London to Pembrokeshire to collect them – though they are such amazing bees that it would be worth it.

Life in the balance
This year, most of my losses have come about because of my choosing to use fewer chemicals. While I'm pleased with this organic approach, the weaker colonies with lower resistance have not survived. Some would argue that this is no bad thing as it's best to have strong disease-tolerant bees, but it's always grim to lose any.

I hate to see my bees suffering. It causes me pain and sadness, and that's when I realise that my relationship with them goes beyond their being my livelihood. Though they might not be aware of it, I think of myself as so much more than just their keeper. I build up such a close connection with them that to sacrifice any of the weaker bees for the greater good is a difficult thing to accept.

But at this time of year, I have to toughen up because April can be a hazardous month for bees. Their existence is on a very delicate balance. Even if you inspect them once a week, you may find they are fine one week, but the next they are half-starved.

Although there is a cosmic-coloured spread of nectar, as pollen becomes available and blossoms break out to provide early pickings, bees are still at their most vulnerable. They can starve in a matter of days should poor weather set in. If temperatures drop and wet weather descends, bees will hunker down. Some will fly out of the hive but most just sit inside and become frustrated. They brush up against each other more closely than they should, which means that disease is easily spread.

The nest is also in a rapid state of expansion. The bees feast on huge quantities of honey stores or syrup daily as they become more active. Young workers are out foraging, queens are busy laying eggs, whilst nurse bees are brood rearing – feeding larvae and keeping the pupae warm. The large colonies are at even greater risk because they require more stores.

Hive activity can be colossal on warm days as thousands of young bees familiarise themselves with their surroundings. Monitoring the amount of honey is important, otherwise you may find them with their heads plunged into the cells and their bottoms poking out, trying to retrieve every last lick. It's a sad sight when they run out of stores – often there are large numbers of bodies dumped on the hive floor and entrance. In order to monitor this, you don't have to fully open the hive every time. You can probably sneak a peek by peering in the sides when you raise the roof very slightly.

At this time of year, I remove my mouse guards and entrance blocks, which otherwise restrict the bees' goings and comings. It also gives the bees a larger landing pad when they pile in, with pollen nuggets like sweetcorn kernels on their legs. You don't want traffic jams building up outside the hive.

Pollen: a super-food

I want to collect pollen this year. This is something that is not widely attempted in the UK and production happens mainly in Spain, France and Italy from where it is imported into this country. But I think more locally produced British pollen will be popular with allergy sufferers, and in the past few years I have received increasing numbers of requests for pollen to combat hay fever. I am always reticent about claiming that either honey or pollen can cure this anti-social condition; I suffer from it myself but find that taking honey every day has no effect on the amount of snot flowing through my veil in the summer.

Beekeeping with bad hay fever is less than ideal. Weeping eyes and a dribbling, over-sensitive nose that can't be addressed when it is under your bee veil are bad enough, but being stung on the nose by an intruding bee just makes matters worse.

Pollen as a food substance is not used enough here and though its health benefits are colossal, you'll really only find it online and in health-food shops. It contains every imaginable vitamin, while also being known for building strength, brain power and fertility, as a result of which it is considered a super-food by a dedicated few. It is also a taste explosion when sprinkled on muesli or yoghurt.

Specially designed pollen traps need to be fitted to the hive entrances, which brush the pollen off the bees as they go in. My traps – bought on eBay from Eastern Europe – look like tiny porches that bolt onto the fronts of the hives. Each has a slotted screen inside, through which returning bees must pass before they can get back into the hive. This screen pushes pollen from the foraging bees' legs down into a small tray beneath, where it will sit and await collection by the beekeeper.

Once collected, pollen needs to be dried quickly to prevent it from becoming mouldy, as it is still fresh and moist. I use an old honey warmer to do this, but you can also buy specific pollen dryers.

For this reason, pollen traps can only be used in fine weather. More importantly, though, the trap must only be operated for brief periods of time, just a day or so. It's vital that huge amounts of pollen still reach the colony at this time of year because it is fed to the young larvae as a protein nutrient. When I return to retrieve my pilfered pollen I feel like a highwayman.

Some beekeepers mix pollen with honey, yeast and syrup to make little patties which they press on to the frames to give the bees a spring boost. Be cautious though. We've made these before but had everything zapped by a company first to remove any disease spores. 1-kilo bags of fondant and pollen can also be bought commercially now and they are handy for young nucs as they can sit on the top of the frames and don't dry out – the bees clamber inside and nibble away at the treat.

Propolis Propolis, which bees collect from the sticky buds on trees and use for doormats and for plugging gaps in the hive, is also awesome stuff. It has tremendous antiseptic and natural antibiotic qualities. I chew it when I feel a cold coming on but will then be removing it from my teeth for weeks as it is so gooey.

Gourmet honey Chefs are always on the hunt for more unusual, fresh seasonal produce. Honey is a perfect ingredient as it can vary considerably across the season, and also look stunning when presented as chunks of comb. I get called in to Marcus Wareing at the Berkeley Hotel by James Knappett, head chef, to deliver a honey tasting.

The chefs are keen to find a honeycomb that could work on their cheeseboard and I am excited to see what they make of the various honeys that I produce in the capital and around the UK. We end up sampling around eight different varieties but they are keen on the London ones, as their tastes are so unusual and special. In the end East London honeycomb makes it onto the menu.

I'm always keen to hear of new ways to use honey and asked James, a great young chef with some grand plans, for a recipe. It is a little fancy, but if you can find a cream canister you need to try this – you could use it to impress your friends. The guy is a genius, and I only hope that I have enough combs to keep him supplied across the year.

Here's what he emailed me:

Comb Honey with Munster Cheese Foam

Hi Steve,

So my recipe is:

Comb honey, Munster cheese, new potatoes, sandwich cress

I love using London honey with this recipe because after doing a tasting side by side with other lesser honeys you really get to see the complexity of it. It is mellow and not overpowering and goes well with this dish; the gentle blossom flavouring and citrus undertones work so well with the cheese.

Serves 6 as a starter

For the Munster cheese foam:
500g Munster cheese
¼ pint milk
1¼ pint cream
salt and white pepper to taste

Per person:
4 tsp comb honey
6 small new potatoes
Two good pinches sandwich cress

First make the cheese foam:

Warm the milk and cream together then add the cheese. When it has melted, season with the salt and pepper. In a food processor blitz the cheese mixture then strain through a fine sieve. Put in a gas cream canister, add two gas chargers, shake well, set aside. The sauce will be looser if you don't have this fancy machine – but just as tasty.

Cook the new potatoes till tender and keep warm. In a bowl put the 4 teaspoons of honeycomb in separate sections, then add the warm potatoes. Now with the foam canister, squirt on the cheese foam to cover the honey and potato. Lastly sprinkle on the sandwich cress.

Let me know what you think.

Cheers, James

Am I ready? After months of rest when the bees have been safely indoors, I begin to worry about their survival. Although April is the start of a renewed love affair, it's also at this time of year when the sleepless nights set in. I wonder whether I have made sufficient plans for the season, whether ample equipment has been built and ordered for the months ahead. I curse myself for not doing more during the closed months.

The season is already more advanced in the capital since it has its own fabulous microclimate. The daffodils are brown and crestfallen, while in Shropshire they are still spritely. If the weather remains fine all through the spring there may be some early honey crops coming in, though most beekeepers leave this to be consumed by the bees rather than harvest it.

It feels as though the seasons are getting earlier across the South-East every year. My barometer used to be the horse chestnut tree where I parked my truck in South London. If the pink flowers were about to burst, I would race around and put all my honey boxes on the hives as I knew the first crop was on its way.

Unfortunately this mighty tree has since been chopped down by the council during the redevelopment of Bermondsey so I have to look elsewhere for my signal. In order to judge it I now use the local parks that I pass through daily. They mostly have a good mix of trees – willows, cherries, chestnuts and limes are all good seasonal indicators.

An early season is generally a good thing so long as both beekeeper and bees are ready for it and have built up their strength, otherwise there's a chance you might miss the best flow of honey. This year there is unheard-of consistently warm weather, with summer temperatures. At the start of the month it's encouraging but as the weeks of dryness continue, it's a worry as there is little nectar arriving in the hive. Flowering plants and trees seem to be blossoming and then going over really quickly, and the result is little or no nectar – the heat is mopping up any potential crop. I can't believe I am saying this ... but for God's sake rain....

Getting ready for honey

Honey boxes By the end of the month, the good weather is usually settled enough for you to put on your honey boxes. You might not see honey coming in for several weeks, but so long as you're not chilling the hives before the weather has warmed – in which case you can be doing more damage than good – there's nothing wrong with mounting your boxes and hoping for the best. Wait for prolonged good weather – in my case this means a daily obsession with forecasts. I always look at more than one weather forecast a day.

Queen excluders An important little bit of kit is a queen excluder, a mesh made from plastic or metal that prevents her royal chubbiness from entering the honey store and depositing her eggs in what should be cells for honey. She needs to be kept in the brood rearing area.

Inspecting your brood

Regular brood inspections are vital now. When I'm in the zone, even though I know I need to be extremely thorough at this time of year to check everything is OK, I can rattle through one very quickly.

Each frame within the brood box is examined with care and where bees cluster, I gently blow on their backs to make them part – bees are good at hiding queen cells. As a general rule they tend to build them on the ends or sides of the frames.

If for any reason I'm unusually worried about anything I will gently but sharply knock the bees into the brood box before removing it. This will leave just the young bees clinging to the frame – and sometimes the queen who has particularly

81

grippy feet. This technique is used by bee inspectors, who are keen to eye virtually every cell for possible disease, but it's not advisable for the beginner – it isn't good practice to be knocking the queen from the frame and the bees won't thank you for this turmoil.

How to do a basic brood inspection

Five things to check for:

• Is the queen present and laying properly?
You need to look for young eggs standing up in the bottoms of the cells; she should be laying a large number of eggs by now. White and like thin grains of rice, they will be hard to spot at first until you get your eye in, but tilting the comb into the light can help. If the frames are older and darker, a top tip is to use a torch to illuminate the bottoms of the cells – but take care as you'll need three hands. When you spot them, freshly laid eggs should be bolt upright and in singles. The bees will often stop a queen laying a few days before they swarm, so even if you don't find any eggs it doesn't necessarily mean that she is absent. But obviously spotting them is a good reassuring sign – fresh eggs means queen in lay.

• Is the colony healthy?
There are so many viruses, parasites and brood diseases that can trouble the poor bees, and you need to check if they look healthy. This one is harder for the beginner, but there are a few worrying signs to look out for. Are your bees crawling around on the floor, lying on their backs or looking sick? Shaking or just looking weary? Many years ago I was convinced I had some sort of brood disease in my hive, so I called the local bee inspector – he advised me that the huge

My paternal grandparents, probably in the 1930s, tending bees on their smallholding, Providence, in North Shropshire.

David Graves, a pioneer of guerilla urban beekeeping, taken back in the autumn of 2001 when bees were still considered wild animals in New York.

Brushing bees from combs of lime honey with an old T-shirt at twenty-six storeys high, near Union Square.

Removing a honey super, strapped up with rope, from his bees. It will hang from his waist as he descends the ladder, after which it will be wrapped in a black bin bag before being transported on the subway.

New York honey on sale at Union Square farmer's market.

Natalie Hodgson and her village of bees at Lavender Farm near Bridgnorth. I visited her when I was a photographic student in the mid-1980s, and again when I moved back to Shropshire to manage bees commercially.

David Wainwright with his bees in the late spring.

David's picturesque valley in West Wales on the perfect day. Brightly coloured nuc boxes on stands form part of his breeding station.

My first London hive, on the rooftop of my block of flats. The bees prospered and produced a butterscotch honey in the first year before swarming down on to the neighbouring school fence in year two.

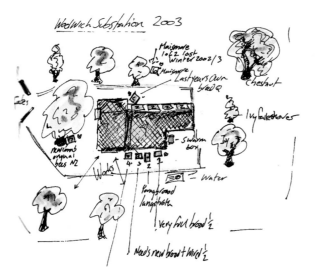

A rough sketch from my bee record book of the electricity substation in Greenwich, a few weeks before the hives were burnt to the ground by vandals.

Our first pot of London honey, in a Kilner jar with a luggage label.

My first bee truck – a hand-painted Morris Austin pick-up, great for moving bees around the capital and going to markets, but sold when Ned arrived as it was an impractical family car.

My first stall at the gates to Spitalfields, before the market was redeveloped with luxury shops, with an observation hive bolted to the tailgate of the pick-up.

Life on the road.

David's tower of honey, perfectly displayed and backlit. Who could resist a pot?

Freshly cut London combs, each with a luggage-label tag and bee stamp.

brown deposits on the ends of the frames were more likely to be the propolis that bees can collect from trees than anything sinister. This was embarrassing but he was happy I had made contact with him.

Do they have enough room?
It sounds daft but space is very important. Too little room in the nest and they will want to swarm quickly. Is there sufficient room still for the queen to lay, or is it all packed with honey and young bees? If it is you need to make space for her. Either add more foundation wax frames and move the honey out, or add another brood box on top or even a honey box. If they're very tight on space, the most effective thing is to split the hive nest frames to make a new colony.

Do they have enough stores?
The other extreme from finding the nest crowded with too much honey is finding it has too little. When you check each frame, you will be able to see what stores the bees have available – often at this time of year I find colonies that have just a spit of stored honey in the corners of their brood frames. They are on a knife's edge waiting for the first bulk release of nectar and they can starve quickly if nothing arrives. If the weather is poor they will soon need feeding. It's another key reason to check.

Are there are any queen cells?
This is a complicated business, but also an essential one. If you don't check, and miss seeing these odd-looking cells in your hive, you could end up with a swarm on your hands. How to spot them and deal with them is discussed in a section of its own, 'Queen cells', just below.

When you've checked your brood, but before you shut up the hive, do check what you need to bring for your next inspection. Feed, more frames or honey boxes – there are so many items and it's why you need to keep notes. Bees need regular attention from now until mid-summer, and the last thing I need is to keep travelling back to sites when I forget items – which of course I do as my head becomes blocked with bee worries and traumas.

Looking after your queen

Queen cells At this time of year you may see queen cells appearing in your hive. Very occasionally, if the weather is unusually warm, or in the London microclimate, I have seen them in late March, but April is when you really need to begin your vigil. These cells are alarming at first sight, not least because they are sinister looking, resembling the wrinkly crooked fingers of old ladies, and they could send the inexperienced beekeeper into a complete panic. You may spot as many as a dozen in a hive – I have even seen twenty once – surrounded by bees that crowd around them to maintain their temperature. Due to the way bees cluster and gather on combs, it's easy for a beekeeper to fail to spot them – another reason to be fastidious with hive inspections.

If you discover them on your frames, first of all, mark the frame with a little cross using your hive tool. This rudimentary bit of flat metal is a crucial piece of kit. But before you panic, take a moment. Close your hive up and head indoors for a cup of tea, and a chat over the phone with your bee mentor or local bee association about your next course of action.

The existence of queen cells can mean many things. Yes, it could mean your bees are considering swarming, but equally it could mean they are looking at making a new queen, which we call super-seizure. This occurs when the bees feel the queen is lagging and about to start failing in her role as an egg-laying machine. If this is the case the queen cells are usually in smaller numbers, and with experience you will be able to tell the difference between a hive that is about to swarm and one that is trying to overthrow the monarchy.

Don't be tempted to interfere until you feel you fully understand what is taking place within the hive. A queen cell takes sixteen days to develop anyway so there is usually time to consider your options. A general rule is that the bees will look to swarm after the cell has been sealed for nine days.

You could give your bees instant space by adding an empty brood box or super and moving the queen excluder above them – this will give the queen more room to lay. If queen cells are going to be raised it's more than likely it will be between the two boxes, making finding them very easy. But the chances are that the bees will want to swarm anyway. Your best bet at this point may be to encourage artificial swarming, a tricky business which I wouldn't recommend to the beginner. This is discussed more fully on page 109 of my chapter on May, super-swarming season.

Queens and breeding
Another common occurrence in April is that of failing queens. At her peak egg-laying capacity, a queen will lay 1,500 eggs a day but sometimes she just gets worn out. If you open your hive and see that the queen is laying five eggs in one cell, a random pattern of egg laying called peppering, it can mean she's losing her mojo. You may also spot an increase in male (unfertilised) larvae which shows that the queen has run out of stored sperm from her mating flight. If

she is becoming infertile the colony will falter. Bee numbers can go down significantly and in my case this could hinder my plan for honey dominance in the capital.

Commercial beekeepers generally replace queens every few years to keep stocks strong and virile; I even know one who diligently changes his queens every season.

If you see a failing queen then I'm afraid euthanasia is required. This is always sad, but there is no retirement centre for failed queens. If a queen is shooting blanks, she needs to be replaced for the long-term survival of the colony. You do need to make sure before introducing a new queen, however, that there are sufficient young workers in the colony still and that it has not become overpowered by lazy male bees.

I used to think you would never see more than one queen in a hive, but last year I saw two co-existing quite happily – perhaps one was the original and the other a young virgin. The old queen's subjects were tolerating this new virile queen – perhaps this would only last until she mated, when her pheromone would be so powerful that the bees would remove the older queen. It was an exceptional case with the bees showing what seemed like respect for the old monarch.

Occasionally you will open your hive to find no queen at all. Without a queen, the bees will be grumpy. You'll spot this as soon as the roof is lifted and no amount of calmness will quell their disapproval. I notice amongst my own bees this year the roaring that you get when a hive has become queen-less – it will happen with a low percentage of my stock, but of course it will be higher if you have older queens.

If one hive becomes queenless, the easiest way to make a new queen is by putting in a frame of eggs from the other hive and allowing the bees to build one up from the existing larvae. This is sometimes called a scrub queen, or an emergency queen. She might be a little smaller and mildly inferior to a properly bred queen, but she'll do the job and see you through the season until you can get hold of another one.

These rather cumbersome lumbering royals can easily fall from a frame when it is inverted and care must be taken with frame manoeuvre. It is hard to describe, but you lift the frame up from the hive vertically, examine one side and basically twist the frame around by dropping one end – usually the left – and slowly rotating it on the corner so the bottom becomes the new top and the reverse side is now facing. This prevents the frame from being horizontal and is the correct way to examine it.

I managed to squash a breeder queen last year with my size 10 feet. I was devastated. She had mothered millions of siblings, as well as having a fantastic temperament. It was such an undignified end to a life of fantastic devotion. I accidently trod on her when she fell from a brood frame I was examining. Immediately, I found bees clustering at my feet looking for her, attracted by her pheromone – but sadly I had already trampled her into tufted grass.

Different strains of bee

In the past, a solution to finding yourself queenless was to import queens from Europe and Australasia. These bees, I believe, are not suited to our miserable weather. More often than not these bees from hotter countries are golden in colour and produce offspring of a similar hue. Traditionally our bees are dark as they are designed to absorb any available sunlight and prosper in damp conditions.

How do I know this? I have tried several strains. On damp days, the dark native bees head out to work while the golden bees stay inside and eat their stores; and when there is no nectar flow, they can be seen robbing from weaker colonies. In other words, they are lazy individuals and not in my opinion good for hard work or ready for our often soggy climate. Their only positive characteristic is that they have a good temperament. But as a bee-master once said to me, 'You want bees that are going to work, son, not hug you.'

There is another problem – and one you need to be aware of – which is that a bee native to the UK does not really exist. Or only somewhere remote like the Outer Hebrides. Sadly, all of our wild native stocks were wiped out by disease about thirty years ago. What exists here is more commonly known as a mongrel hardy bee. I mainly use a variety bred by my dear friend David in West Wales, who set up a breeding programme a year or so ago in his damp chilly valley. His bees prosper when they are moved south to warmer climes as they take advantage of the bounty of food sources. As they are hardy bees, they are suitable for high urban rooftops. Peter's more gentle bees are good for places where they're likely to meet people. Some strains are really tough and resilient against diseases and British weather. Hardy bees are great because they're hard workers. By comparison, some of the lazy gentle bees I've worked with – like the golden warm-weather bees – were so soppy, you could almost hug them, and they certainly didn't require you to wear a veil or protective clothes. I prefer a bee that is harder to handle and a little more feisty.

These are all characteristics you need to consider when investing in your first bees. The best advice I can give you is never just get one colony. Of course you might only have enough room for one hive on your veranda, balcony or rooftop, but if at all possible, I would encourage new beekeepers to have two hives. The main advantage of this is having a spare hive to exploit, should the other suffer any problems.

Smoking your bees

Back at my new plot of land in North London, the hives are
settled, the bees seem to be relishing their new warm location
and I've caught up on some sleep. I'm considering Peter's
rickety old bee shed. It will make a crude store for equipment
but luckily you don't need that much space to house bee
gear, even for several hives. Even if, like me, you're a real
hoarder and keep everything, you can still manage with the
smallest of sheds. However, you will need space in the winter
for honey boxes and a good smoker and fuel.

A smoker is a crucial bit of kit in the beekeeper's arsenal.
Used correctly it gently subdues the bees, so you're less likely
to be stung. It also causes them to gorge themselves on stores
and in turn they become sleepy – with their bellies full, they
can also not curl their abdomens round to sting. If you over-
smoke bees, however, you will send them into a panic; they
become grumpy and race around, generally making your job
more difficult.

The fuel you use to create the smoke is crucial. You need
to burn something that will take some extinguishing and that
is easy on the nose. Peter's bees here loved the smell of his
pipe tobacco. At the moment I'm burning rotten wood from
Shropshire collected by my friend Kingy the gamekeeper,
which sits in my truck in an old pheasant-feed bag.

David has settled on leylandii hedge trimmings – this ever-
green when dried in large trays in his huge honey warmers
smells delightful. It is less aggressive than some of the things
he has experimented with in the past, including string, loo
rolls, dried horse and cow shit, puffballs and even his old
clothes. All these burn but their smell … well, you can imagine. Sorry, David. I prefer pine cones and dried bark. In
Greece they use dried thyme and rosemary twigs, more

romantic and aromatic, but not practical here unless you have an abundance of herbs.

Smokers come in a variety of shapes, designs and sizes. Last year I managed to get through three new ones. They often get driven over, left behind, fly out of the back of the truck or just fall apart. Quite a few come from Eastern Europe and you need to consider their design to make sure you get a good-quality one. I like a fair-sized chamber that will enable the fuel to smoulder for some time. The bellows are usually the first thing to perish, but these can be replaced. I also de-coke them weekly as they burn better when the thick tar is removed from inside. A cage around the chamber prevents you from burning yourself, especially if like me you wedge it between your knees to leave both hands free.

Make sure you keep spare fuel by you. Running out of smoke during an important inspection could result in you wandering off to find some more, leaving a hive open for five minutes. The bees will not appreciate this and will let you know upon your return.

Smoking bees is an art; it takes practice. It's a bit like when someone new to scuba diving breathes in heavily on the air tank. Beginners think huge clouds of smoke will prevent them from being stung. In fact, less is more. Huge plumes of smoke will harm the bees, but use too little and they can bubble over the frames and any manipulation becomes impossible. Sometimes I use no smoke at all. With certain calm and gentle strains, you can feel and sense when these tactile creatures are not happy with your presence.

So if now as a beginner you have your glistening new smoker and something to burn, I would suggest you have a quick practice before your bees arrive. More importantly, make sure you can put out the smoking gun when you need to. Extinguishing the beast is important, especially if your hives are on rooftops. On urban sites I have fire buckets full of sand with metal lids. I also plug the smoker's spout with a shaped cork, which stops smoke alarms being set off as I descend to street level. Sometimes I use moss or tufts of grass twisted into the funnel.

When working commercially, I often leave the smoker burning all day, hanging from the back of my truck. If the balance of fuel is right, they don't smoke too much when left, but a few puffs from the bellows and they ignite again. Lighting from scratch can be tricky, especially in the wet. I usually find somewhere sheltered and start with a dry piece of toilet paper or tissue. Then when I have a good flame, I add my fuel, puffing gently. Remember, you don't want a flame-thrower, just a gentle plume.

From spring to summer Beginner beekeepers will now be starting to look forward to fetching their bees and generally having a happy time getting to know them over the summer. My grandfather used to say that beekeeping was relatively easy in your first year – it is unlikely that young bees will want to swarm. For the rest of us, however, the approach of May – prime swarming season – brings with it some sense of trepidation.

Of course, there are those who believe that swarming is a good thing for bees. It breaks the natural cycle of the Varroa mite as brood rearing stops. The Natural Beekeeping Trust shares the view that it is a natural phenomenon that ensures the bees' survival. While this is true enough, if you are looking after bees in London, there are other things to consider, such as the impact swarming may have on people, many of whom panic at the sight. You simply can't allow it to happen in crowded urban areas – or at least, you should try to prevent it. I would quickly get a terrible reputation among my fellow beekeepers, who would probably end up being deployed around the city to mop up after my incompetence. As you will discover in the next few chapters, I always do my best to prevent my bees from swarming but that doesn't mean it never happens.

By the close of April, my dancing has worked and the rain arrives. It's been a long wait and the ground is heavily cracked at the North London site, but the water stations I have built for the bees are now overflowing when I arrive and of course those Welsh bees are flying. They are hardy types – good girls. Bring on troublesome May – I'm ready for it, with all its traumas, exhaustions and relationship breakdowns. But my devotion continues for a harmonious bee business.

What potential new beekeepers should be doing:

- By now you should be in touch with breeders about when you might be able to collect your new colony – usually June.
- If you haven't already done so, join your local beekeeping association – they can give support and mentoring, plus a degree of insurance.

More extensive beekeeping tips:

- Remove mouse guards and entrance blocks from your hives.
- Start doing regular brood inspections and keep an eye out for any early queen cells.

- If you need to get any over-wintered queens then give your breeder a call.
- Assess whether your bees need an energy boost and if they do then feed them some syrup.
- For those few intending to collect pollen from the bees (a complicated business) now's the time to do it.
- Keep obsessively checking the forecast for warmer weather. When it arrives, put on the honey boxes and queen excluders.

MAY

MANIC MAY

For most people, May is traditionally a month of fun, merriment and celebration. Summer is here at last. For me, it signals a relentless period of hellfire and sleepless nights. For this is the month of the swarm. My every move across the capital is controlled by the behaviour of my bees and I dash frantically from site to site checking that none of them look as if they're about to take flight.

I work late; I forget to eat – even a boiled egg seems to take up too much precious time. Several times a week I crash out in my truck, or I sleep under a tarpaulin in the wood at the North London site, or even on a rooftop. It's the only way I can get everything done. But this itinerant lifestyle doesn't often reward me with a good night's sleep, which is what I really need in this frenetic month.

Consistently warm weather in May can signal amazing London honey. With the right conditions, bees produce a dark honey from hawthorns and later blackthorns, as well as chestnuts and sycamores. The most successful hives can create half a box of honey in the first week.

If you look in your hive just a week after seeing such promising early supplies, you may wonder who the devil has stolen your honey. I don't worry if this happens; the nest will be reaching a colossal size and there are lots of hungry mouths to feed. Besides, the honey does actually belong to the bees. If a week of sunshine is followed by one of bad weather, chances are the honey will be quickly depleted. The bees will have to stay inside, munching their supplies rather than bringing in more.

This year, the madness starts earlier than normal thanks to the unusually warm April which enhanced the bees' development. Everything flowered at once, bringing the cycle forward by several weeks. Although an early nectar flow sounds like a good thing, warm weather can mean that the flowering stage – which provides nectar – is short. Sometimes the nectar is dried up by the sun before the bees can get to it. That rain at the end of April arrived just in time.

Moisture is needed to offer longevity to the nectar flow, and sometimes the window of opportunity for bees to bring in a decent honey crop is very narrow. Meanwhile, readying them for this bonanza period is the crucial role played by beekeepers; it's a delicate art, requiring patience, planning and some good luck.

Bees must be in the right condition to make the most of good weather. Ideally, you want your hives to be bursting with bees when you lift the roof off, but often numbers are still low if the flowering season comes early and the hive is recovering from winter losses. If you open up your hive during the nectar flow and you don't see bees bubbling out of the honey boxes, then it is unlikely you will get a good crop.

It's a delicate balance. On the one hand, you want a colony big enough to bring in honey, but on the other a hive that is packed to the rafters is also more likely to swarm. The challenge is to produce the largest, most productive colony possible without triggering the swarming instinct.

Last year, my honey crop was impressive because my bees were young and virile. I'd invested in them the previous year so I was rewarded with lots of honey. The drawback was that I was so busy harvesting honey I didn't have time to work on making new colonies that would be productive this year.

I'm now reaping the consequences, dealing with several depleted colonies that need building up to ensure next year's honey flow will be better. It's quite normal to have one good year, followed by a poor one, especially given all the environmental factors beyond a beekeeper's control. As a rule of thumb, a good honey crop is a mixture of ideal weather conditions, careful planning and good management. It is unlikely you will produce a consistent crop across the three seasons – spring, summer and autumn – and at best I usually manage two.

All of this explains my current obsession with the weather. David's got a fancy tool to help him monitor it and as soon as I can get myself sorted with a bee HQ in London I'm going to get one too. Installed on David's roof is a weather unit on a giant pole which looks like the mast for a pirate radio station. It feeds weather reports remotely onto a screen on his desk and when I've been there I've scrutinised them, spellbound. For most beekeepers, local weather reports will do the same, but they vary considerably. I've decided that I too will invest in a more accurate method this year.

Despite my manic checking of the forecasts, I still find myself disastrously unprepared for the hottest day of the year so far. I end up stuck in gridlocked traffic facing a neon sign that tells me the Rotherhithe Tunnel is shut. London Radio explains that there is a car fire inside the smoggy tunnel. My van's temperature gauge is close to popping and all of the windows are down.

I bought the Berlingo for deliveries around London. Thanks to its high clearance it's easier on my back when I'm loading and because it's enclosed the honey inside is safe from marauding bees. I can even sleep in it. On board today is a huge colony of bees that I collected last night from a new site in East London. At this time of year a mature colony can consist of more than 50,000 bees, and being transported like this can have them gasping for air. They need to be kept cool with as much air as possible directed underneath or above them to prevent meltdown.

No matter how experienced or knowledgeable you are, you still need to work hard to minimise stress when you move bees. As with humans, stress can induce in them numerous diseases and ailments, and the bees' welfare should always be paramount. Smell and noise are two things to monitor, as they should alert you to when all is not well.

Dead bees smell like a rotting animal, and if things are getting bad the buzzing noise will grow more intense. The bees will be frantically fanning their tiny wings to try and cool themselves and their combs before apocalyptic melt-down ensues.

When bees overheat and suffocate it is not a pretty sight, one which I hope you will never see. I have witnessed it twice. The worst was on the way to the heather moors in Wales. To this day, I'm not sure exactly what I did, but I suspect that I went over a bump too quickly and the bees fell to the bottoms of their hives blocking the air vents. When I opened the hives, I was confronted by a mash of stinking melted wax with thousands of tiny bodies littering the sticky goo – a tragic sight. There are no survivors if you let your hive overheat.

Today, the bees on board my truck are being taken from a new site in East London to another one in Hackney. They had swarmed in East London and I always prefer to take bees to a new location after swarming because they settle down more quickly; it also stops them from clustering around the area where they swarmed. If you do this it's important to give them a good feed of syrup to settle them in – a sort of house-warming present – as their only stores will have been in their bellies for the journey.

The morning starts badly when I oversleep. I'm relishing the friend's sofa bed I am camping on for a couple of nights, and the comfort it brings, but I also know that it is dangerous for my body to relax completely at this crucial part of the season. If I do I feel as though I might not move for weeks.

The number of hives I manage is now expanding across the capital. With new sites and customers turning up weekly, I'm forced to work later into the night. This morning's over-sleeping means I'm now moving my bees across London later in the day than I would have liked, risking the lives of my passengers in prime London traffic.

I can feel myself starting to simmer. The animated buzzing from behind me inside the van refuses to be drowned out by Radio 1's garish breakfast show. Could this be the first time I lose bees through suffocation in London? Dear God, I hope not.

The reason I am overtired and running late this morning is due to a typically frantic May day yesterday – exhausting and perilous. I spent it hunting down this lot of absconding bees and checking on other colonies that had been looking as though they might swarm.

Spitalfields Market The day had begun bright and early when I collected a new volunteer from outside Old Spitalfields Market. By now, my volunteers are mostly trained up and battle-ready, prepared for any skirmishes that may arise. But as luck would have it, yesterday's assistant was an inexperienced one. Áine is an out-of-work Irish actress from Hackney who works in my local coffee house (which stocks my honey). She makes up for her lack of experience with bags of enthusiasm, and I soon became confident that she'd pick things up quickly.

Typically, I was too rushed for breakfast – a bad habit of mine. Sometimes there's just so much to do at this time of year that I forget. On long bee days, eating something first thing is crucial. I've been caught out before when I haven't had breakfast. Like a bee on a sunny March morning that can't summon the energy to get back to its hive, I start the day with gusto, then run out of steam mid-morning. This job is so demanding on my strength; it saps my energy reserves faster than I anticipate, and at times I have found myself on the verge of collapse.

Having learnt my lesson, I generally try to make breakfast a priority. My favourite only takes a minute to prepare. I take a large pot of natural yogurt, tip half into a bowl and put that back in the fridge. Into the pot I then spoon muesli, half a banana and of course some London honey. This sits nicely in the truck's cup holder and I can dive into it when I'm sitting in traffic jams. It provides slow-release energy, imperative on these adventurous days.

Yesterday, having failed on the breakfast front, I rushed across town to meet Áine outside Gardner's – a famous market trader's store selling everything from environmentally unfriendly plastic bags to giant balls of string. It's a trader's dream – they even have those vibrant bits of star-shaped card that you scrawl your special-offer prices onto.

Spitalfields was the first market I sold honey at, back in 2000. The market manager was a charismatic rocker called Eric with long blond hair that cascaded down his back. He was so supportive of my bee mission, he often forgot about charging me rent.

I would park by the entrance gate next to Jim Juice, maker of fantastic smoothies for the hungover grazers, while we made our stall up in the back of the pick-up, often strapping an observation hive onto the tailgate to draw in the crowds.

Exquisite French Kilner jars sourced from a local supplier (and brought back by customers for recycling) were full with our bee's liquid gold from hives in Bermondsey and Plump- stead. This was when I first decided to use luggage tags and string for labels, the backbone of my company's labelling today. It looked (and still does look) simple and stylish.

As a first introduction to market life, it was fun to dip in and out of being a trader – a far cry from my committed approach to sales at markets today. We gave away most of the honey and chatted to hundreds of people about the virtues of keeping bees in London. Looking back, I realise we were sowing the seeds of the London Honey Company.

Since its redevelopment, Spitalfields has become very different. Flaunting high-street shops and restaurants, it has a polished feel. Half the ancient buildings have been pulled down to create offices. It's not so much about smaller traders making their first steps with new products any more. If you ask me, it has lost most of its soul and colour.

Áine and Esther Áine and I headed straight from the market to the North London site. After a quick cuppa made on the camping stove, we set about checking on the bees. We were soon shedding several pints of sweat in our bee suits and when the temperature at last cools, we load several hives onto the truck and head off to restock Fortnum's bees.

We're bringing in some new hives because numbers are down on the department store. I lost so many bees over the winter, I'm keen to build up these lucrative hives. We move the new bees with care onto the rooftop in special mesh bags, designed, like my syrup-making apron, by Tarpaulin Mike. His bags keep the bees ventilated, while also preventing any escapees.

That done, we pick up a swarm of my errant bees from a shop front in the East End and finally, as the sun sets, we finish the day at a new site nearby, putting more honey boxes on the hives. After such hard work, I wondered whether Áine would volunteer her services again. It sounds brutal, but a challenging day like this is often the best way to start volunteers and test their bee devotion. If they offer to help again, I know they really want to be involved.

In many ways, Áine passed her bee initiation with flying colours. Not only did she bring along some fine honey cake (turning up with cake is not yet a requirement for volunteers but I'm thinking about it) but she also demonstrated the ability to remain unbelievably calm, meaning she received not one single sting, despite wearing only a smock and veil and thin socks much of the time. Some people are just naturals; it's not something I can teach.

Last spring, another actress – who attended my winter course and was following it up with my practical introduction class in the spring – began completely freaking out when I opened the first hive. Beginners usually know roughly what to expect after my classroom courses so I was a little alarmed at her reaction.

Esther is a Northern lass with a broad accent and she started screaming, 'I've been stung on my 'anny,' as the bees circled around us. None of us knew what she meant until she started ripping off her overalls and waddling back towards the bee shed. A bee had crawled up inside her trouser leg and stung her between the legs – we realised now that she'd been dropping the f. Bigger pants were required next time, I suggested – but tucking your trousers into your thick socks is a better tip.

The hazards of swarming

Swarming and how to prevent it is constantly on my mind this month. It dominates my free time like a jealous bunny boiler. Each hive has the possibility to deposit 50,000 bees in a high street or school near you and that is one hell of a responsibility. One queen cell missed, one telltale sign that a swarm is imminent overlooked and my reputation could be in tatters. I wake up several times a night, thinking about specific queens that I know are hell-bent on trouble. Some delinquent majesties, you just know, are set on abdication.

As an urban beekeeper, I am paid to maintain hives for customers and this professional service means I must do everything within my power to reduce the chances that bees will take off and settle somewhere unsuitable. I have a huge responsibility.

The majority of people who come into contact with bees will be petrified of being stung or harassed. It is therefore important to spread the word about how amazing keeping bees is to those around your hives and explain to them the great virtues of their new neighbours. It also helps to provide a little info about what people can do if they see a swarm, and to reassure them that just because bees are swarming it doesn't mean they are going to attack anyone. Generally bees are very calm when they swarm; they are not aggressive. It goes without saying that you should stand at a safe distance, and avoid taking pictures with a flash which could freak them out, but otherwise no one should miss the chance to witness bees swarming. It's an incredible sight, a beautiful, natural phenomenon.

One of the trickiest swarms I have ever collected came from my old flat in Bermondsey. The roof garden had expanded and bizarrely Sarah Beeny from that show *Property Ladder* was filming the newly renovated kitchen inside the flat. It was the cameraman, who'd been having a cheeky fag out of the bedroom window, who first alerted me to the abdication; he came running into the kitchen, screaming that the sky was turning black with bees.

They eventually chose to settle down below on the fence belonging to – of all places – Tower Bridge Primary School. This was a real baptism of fire. Though the bees looked a picture amongst the scraggy clematis that wove in and out of the trellis, they were about 20 feet from the ground and retrieving them was going to be more than a little tricky. I considered contacting Dockhead Fire Station down the road.

I eventually secured a stepladder from the school's caretaker and proceeded to coax the little darlings back into an inverted empty hive. It was a fairly traumatic event, as the film crew, numerous passers-by and all the children were looking on the whole time. I swore that from that day on I would be more meticulous with my swarm checks.

Why do bees swarm?

What exactly is swarming and why do bees feel the need to cause untold woe and trauma for the beekeeper by doing it? Various factors can trigger it, but the key detail to remember is that swarming bees have only one mission – to find a new home.

The purpose of swarming is simply to enable honeybee colonies to reproduce. The old queen bee will gather a large group of worker bees, usually around 60 per cent of the colony; together they will fly off in search of a new place to make their home, leaving the new or emerging queens and the rest of the workers behind in the original hive. That's why the top sign of swarming is normally spotting the development of those unusual-looking queen cells in your hive.

Swarming bees will have gorged tummies of honey for their journey and usually come to settle in a cluster within a few metres of their original hive, from where they send out scouts to find a new secure residence, which can be some distance away. If you're lucky, they will stay at this initial site for a couple of hours or so, if they are not disturbed by noise or people and especially if the weather turns bad. Sometimes, they all decide they want to pop out and do a few laps around a garden or building and then end up returning to the hive from which they departed. Phew!

There are several reasons why your bees might swarm, one being the large size of your colony. I try and keep my colonies as humongous as possible as I know that they are more likely to achieve a substantial crop of honey; but an overcrowded colony with no more space for the queen to lay is also more likely to swarm. Usually, the gamble pays off and with careful manipulations I can generally persuade any bees showing signs of swarming to stay.

Scooping up errant bees

Occasionally, it does all go wrong; and that's why this year I find myself up a lilac tree at a site in South London, complete with a climbing harness, from which hangs my lit smoker, a pruning bush saw and various sling and tapes. I was good at climbing trees as a kid, but somewhere in the back of my mind lurks a childhood mishap, the rowan tree I fell out of in Auntie Pam's back yard in Cambridge.

I need to get within sniffing distance of my massive swarm, so I use a pair of old secateurs to trim off smaller twigs. Besides, the tree looks a little neglected from this angle and needs a quick clip. Then with a quick blow to the branch, I bang the bees easily into the company's staffroom bin – bingo! It's all about technique and confidence.

The age of the queen and the strain of the bees are other contributing factors. Some breeds of bee are more prone to swarming than others. I used to have some Carniolans, a subspecies of the Western honey bee, native to Slovenia, that would swarm on the 3rd of May virtually every year. These bees required tremendous attention to prevent their abduction.

Weather also plays its part and you often find that bees will get incredibly excited on warm sticky days. I used to pray for rainy weather when I was less experienced as I knew my bees would be unlikely to swarm in the wet – a complete contrast to today when I yearn for those clammy days to trigger enormous nectar flows.

Some of my hives are easier to monitor for mass breakout than others. At Fortnum's, the two remote webcams which feed live pictures into my computer, Big Brother style, can give visual hints. At night the light pollution is so great that I can see the bees hanging in clusters on the entrance as each hive becomes crowded. They are no longer able all to fit inside and are frantically trying to cool the hive with intense fanning. When the webcams were installed, the Fortnum's security staff joked that they were better quality than their in-store surveillance ones – although sadly I cannot see the pollen detail on the bees' legs as they fly in. This is another vital indicator – remember, more pollen equals population growth.

Over the past ten years, I have heard of honeybees taking up residence on shop fronts, dustbins, trees, fences and my personal favourite – lampposts. Bees look incredible clustered around a lamppost, although they're a nightmare to remove them from, unlike a tree or fence. You can't easily knock the bees off something so rigid, so you simply have to scoop them up with your hands or use a gentle brush or feather.

My favourite gadget is what I call my defibrillator, a cobbled-together bit of vacuum cleaner, generator and loads of plastic tubes and pipes. It looks like something from the Ghostbusters film set, but it's my bee vacuum, an effective tool for removing a swarm of bees from difficult urban surroundings – it literally hoovers them up. The bees go into a separate swarm box where they sit until you can knock them into a new hive in the evening. You can buy these machines commercially nowadays, but I prefer this old contraption which an aged beekeeper in Surrey showed me how to make ten years ago. Those early experiments did sadly see a few casualties, as the prototype models used over-powerful vacuum cleaners.

Should you ever have to collect a swarm, you will need a suitable ventilated box and some new frames, as well as a white bed sheet for wrapping them up, a soft brush or feather as mentioned above and a hive strap. Oh, and plenty of gaffer tape for any leaks – there will always be leaks! If you have a number of hives, it's a good idea to keep this equipment together where you can lay your hands on it quickly.

Rest assured, May shouldn't be as stressful as all this makes it sound. But whether you have two hives or two hundred, diligence is still required in this unpredictable month. My best advice for a novice would be to remain calm at all times, even if your bees have swarmed, to seek as much advice as possible and never to book any long May holidays.

Believe me, no matter how busy you are, finding time for weekly checks is far less stressful than trying to coax the bees back into a cardboard box in the middle of Oxford Street, surrounded by people taking pictures and asking if you are being stung.

Checking hives for signs of swarming

Each individual hive needs to be carefully examined every week. Even though it is time-consuming, you must be meticulous. Every frame has to be inspected like evidence from a crime scene. The details of how to do this are given in April, under the heading How to do a basic brood inspection (on p. 82), but the main points to remember are:

- Is the queen present and laying properly?
- Is the colony healthy?
- Do they have enough room?
- Do they have enough stores?
- Are there any queen cells?

Bigger hives

As overcrowding can be one of the reasons why bees swarm, giving a queen more space can help minimise it. The standard British hive is very small and I've always thought that bees get easily cramped in it, which is why I choose to use huge brood boxes. I remove frames as the nest expands, allowing the queen more room. I believe my huge boxes are one of the reasons why I'm not troubled more by swarming in London.

When I first considered having bees in the city, I bought a WBC (named after its maker – William Broughton Carr) – a cottage-style beehive. I liked its rustic design and it fitted with my vision of bringing some of the country to London.

It turned out to be an impractical move. The hive was far too elaborate with a fussy design and double walls, which meant it was a hassle to get inside. I soon changed it. I'm surprised that the WBC remains a popular design in the city, though I suppose it is partly for aesthetic reasons. A media company have recently asked me to install six of them on their new building in Euston, overlooking Regent's Park. They will need strapping down so that the roofs don't fly off in the winter. Gusts of wind howl around tall buildings in London, channelled by the urban jungle.

Artificial swarming

If your bees are looking as if they are about to swarm, then you could consider the great manipulation that is the Artificial Swarm. There are various methods, and they depend on what outcome you want. You could manipulate your bees to produce a new queen; alternatively, you could split them and create new nuclei to expand your stocks. This is not something I would recommend to beginners.

I have developed my own crude version to work alongside servicing bees in the busy city. What I generally want to do is make new nucs to replace lost colonies and maintain strong young stock; depending on the strength of the colonies I try and make two fresh nucs from each mother colony. This gives the original queen plenty of space to lull her and the bees into thinking they have already swarmed and need to rebuild their empire. Mother colonies that have been knocked back should still go on to produce a crop; the nucs are less likely to do so until next season.

It is important before you start that you have spare equipment ready and that you are competent at spotting queens and queen cells – if you miss the queen and put her in a nuc box then you're back to square one; likewise if you miss a queen cell on the returned frames. You must be methodical in your inspection – and resist the temptation to rush or, if this is your first time doing it, to panic.

Firstly, I go through the colony and assess the state of play. When I find any queen cells, I use my hive tool to mark the top of the frame with a cross. I don't knock off any cells until I have made an extensive examination and ascertained what is going on in the hive. The bees could just be attempting to replace the old queen. Alternatively, if there are hardly any bees, they could have already swarmed, which means you will need a cell for a new queen.

When (with any luck) I come across the existing queen, I isolate her by putting the frame she's on in an empty nuc box with the entrance closed. Then I pick four to six good frames of brood and knock off any queen cells I can see. You must be diligent here – the frames need to be mainly sealed with no fresh eggs, meaning the bees will not try to raise an emergency queen from them. I split these brood frames between two nuc boxes, into each of which I also put some fresh foundation or sterilised heavy comb stores, and finally a good shake of young bees from the mother colony.

For each new nuc, I will have with me either a queen cage containing a young mated queen and some attendants, or an incubator with a sealed queen cell about 48 hours away from emerging – a virgin.

One or the other is then put into each fresh nuc box with the other bees. If you're using a mated queen, then put the cage with her and a piece of fondant blocking its entrance straight into the box; this will delay her arrival onto the frames, giving the bees time to become acclimatised to her pheromone. A ripe queen cell should be placed, facing downwards, in between the top bars of the brood frames if this is an option.

These boxes are then sealed up and put in a cool place with a breeze. At dusk they will be moved onto a new site further than three miles away, to prevent any bees from returning to the original site, where they will be fed and opened up.

If all goes to plan these nucs will prosper and the virgin queens mate within a few weeks. The colony will build up for the winter and sometimes might go on to produce a small honey crop, but their full production will be the following year.

Meanwhile, back at the mother colony I remove all the remaining queen cells, by pinching them out with my fingers. (You could use these cells to make up your own nucs, if selected at the right stage – sealed and close to emerging. Put in two per nuc – unlike the good strong incubated stock which I've used above, you should have one and a spare. If both hatch successfully they'll battle it out.)

I then return the frame housing the original queen to the hive – centrally if possible, so she is surrounded by one or two frames of brood – plus some new foundation frames to make up numbers and prevent brace comb from being built. New foundation also helps reduce brood disease from old combs and keeps everything fresh and clean. All being well the remaining bees will pull out the new foundation for the old queen to lay in and the reduced colony will carry on as before.

In a good season you might have to perform this operation several times, especially if the queen is young and virile, meaning you end up with loads of young fresh nucs – brilliant for someone like me but not if you only have a tiny balcony for your bees. If this is the case, then in a more complicated version you can eventually reunite the split colonies. But that's a story for another day.

Keep careful records

Another important part of a beekeeper's approach to swarming is keeping careful records. If you've tried one tactic or manipulation to prevent swarming, you want to be sure you remember exactly what you've done and to which hive. Only by keeping track in this way will you work out what effect your action has had. You always think you'll remember it, but of course a week later, with so much going on, the chances are you won't.

As a rule, what I do to one hive in terms of manipulation, I try to do to the others on the same site. This makes things much simpler. If I'm in a hurry I might just sample a few hives to gauge what is going on, but ideally you need to check every hive individually. If not, you'll have to return a few days later to check the rest.

Each of my hives has a small metal plate on the bottom right-hand side where I can attach stickers and notes – David's idea. On it is the age of each queen – this is usually drawn on a coloured sticker (a different colour for each year) – and her breed is also scrawled on top.

Then I keep a logbook with notes for each site. Not only does it have details of manipulations undertaken, it also keeps track of when I last inspected each hive and reminds me what I might need to bring for my next visit. Pruning shears, perhaps, for an overhanging tree or more honey boxes. You can tuck your notebook away under a hive or keep it in your bee shed – whatever works best. Mine sits on the dashboard of the bee truck in the summer and is backed up by an online spreadsheet.

Still stuck in traffic Back in the muggy jam in Rotherhithe, I'm still stationary. I must think quickly to save my colony on board. I could always unload it, abandon the van and hope to find a suitable garden. Once before when my ageing VW Golf broke down, I unloaded the bees into a field on the side of the M1 where they were very happy while we waited for help. In Deptford my options are much more limited.

I rummage around for a water bottle in the passenger footwell and splash water gently through the vented mesh so it runs into the bees' box. Then I make a swift U-turn – sorry, officer – and head for the sanctuary of my mates' garden not far away.

They aren't home. But I reckon I can just about clear their back garden fence even while I'm carrying the hive, and deposit it near the compost bin. Then I can open the entrance block to let the poor things out. Getting a hive over a fence alone is tricky and it tips up a little but I finally manage the whole manoeuvre. I stuff a scribbled note through Doug and Jane's door: 'Hi there, bee emergency, sorry, you have new residents squatting in your garden…. Life or death – will come back and collect later this eve! Steve x'. For now, crisis has been averted.

As it turns out, the bees stay in their new home for a couple of months and even produce some delicious honey there. I secretly hope they might be adopted permanently; but eventually I'm called on by Doug to move them. The colony has grown enormously by the time I pick it up and I'm thrilled that a hive that nearly came to a sticky end on the London roads has managed such a comeback.

The first honey

It's time to consider taking off that light delicate spring honey – demand for it is even greater than usual this year. I try and keep each box separate as I think people appreciate the regionality of all the different boroughs. It gives customers the opportunity to purchase local honey and we label it up accordingly, this year with the first part of the postcode stamped on the lid of each jar: SE1 for Bermondsey and so on.

The following recipe for honey with mint and peas is simple and the most important thing in it is fresh spring honey. Try it – it's totally lush.

It was given to me by the aptly named Tom Bean and it comes from his eccentric side. I first met Tom ten years ago through David in Wales. He is a wild forager and comes just for the crack to help with our bee moves in the autumn. I absolutely relish his company and the opportunity to forage for some more unusual tucker on our trips; plus he is very good at keeping me awake with his banter when we are on the road. Last year I devised a quiz for our trip back from the heather, with such questions as name five famous Barbaras, or name five Müller Fruit Corner varieties as we pass their huge factory in Shropshire. Mind-bending questions I know … almost home, Bees.

I Eat My Peas with Honey

I eat my peas with honey;
I've done it all my life.
It makes the peas taste funny,
But it keeps them on the knife.

Tom Bean's Honey with Mint and Peas

mint leaves – fresh
springtime honey – first of the year
peas in their pods – raw, small sweet peas

Carefully dribble a line of honey down the vein on the top of the mint leaf. Open a pea-pod and transfer the peas to the leaf, sticking them down the line of honey. Gobble as you go or make a plateful and present to guests as marvellous, refreshing, appetising canapés.

We've survived May closes, with the majority of bees still in their original homes. My juggling and sorcery can be classed as an early victory, although I do have to wrestle another swarm in East London that has stuck itself to a Victorian wall. I have never seen anything like it – it looks as if the bees have just been thrown at the brickwork. It's the easiest swarm to collect, as with a quick brush it's housed in seconds. But a resident shouts from a neighbouring balcony that it was the most alarming thing he has ever seen.... Whoops. Well, he appears to be smiling still, so I whip them quickly up to North London on my Vespa.

What potential new beekeepers should be doing:

- It's countdown to the arrival of your new bees in June. Triple-check that absolutely everything is ready for their arrival.

More extensive tips for the established beekeeper:

- Keep doing regular hive checks. It's no good taking any chances – be diligent for signs of swarming.
- Best not to take any holiday for the next few weeks and to answer all the calls on your mobile in case there's a swarm.
- Have some key bits of equipment – such as a white cotton bed sheet and a mesh-ventilated box – by the front door in case you're called out to deal with troublesome bees.
- Have extra frames of fresh foundation made up and a spare hive or two – to house splits and swarms. The last thing you want is to be banging together equipment in a hurry.
- If the conditions are right, and you're an experienced hand, you may want to consider artificial swarming as a method of swarm control.
- Keep careful records of all manipulations performed.
- If it's ready, take off the first lovely spring honey.

JUNE

PROGRAMMING THE SAT NAV

On warm sunny days beekeepers will be noticing increased activity at the fronts of their hives. The growing bee traffic is the result of a population explosion. The queen has been busily laying eggs and the colony is booming. Egg-laying usually reaches its peak on the longest day of the year – 21st June – although since the eggs take three weeks to hatch, it will be in mid-July that the hive achieves its greatest population.

Seeing such a crowd of bees outside a hive can be a little terrifying. To a novice beekeeper, it can look as if they might be contemplating swarming. But don't worry, it's perfectly normal. The newly emerged bees are simply familiarising themselves with their front door's position and getting to know their surroundings. This is when the bees perform what is called the 'midday flight', their first trip outside the hive, in a process which programmes their sat nav to home. They will soon be embarking on foraging expeditions and in my bees' case circumnavigating the capital.

In the countryside, it is believed that bees can fly up to a maximum of 5 kilometres on their foraging trips, returning to the hive with their booty several times a day. They refuel, direct other bees to the best source of nectar and set out again on their relentless harvest.

In urban areas, I feel this distance is reduced, mainly because architecture gets in the way. They have to dodge high-rises, cranes, radio masts and church spires, which for the tiny creatures creates a much more arduous journey. They rarely drop down to street level to buzz along bus lanes and cycle paths, but instead head up above the rooftops, spiralling down to forage when something sweet takes their fancy. At which point they have the traffic to contend with.

Observing how bees navigate In my second year as an urban beekeeper living in Bermondsey, I learnt about bees' navigational behaviour the hard way – when they swarmed on to that school fence. After this I decided it would be helpful if I set up an observation hive inside my flat. I wanted to learn more about what went on inside the hive, if nothing else to prevent the drama of collecting urban swarms.

Observation hives are remarkable things, addictive to watch. Instead of hive walls, there is glass, which allows the viewer true insight into what takes place within the usually dark brood nest. The huge glass cabinet enables you to watch the bees as they go about their daily duties. Although in the short term being the object of such close scrutiny doesn't seem to bother the bees, because they prefer to be in the dark it's best not to keep the same bees like this for more than a few months and it's definitely unfair to keep them in it over winter.

On warm days, an observation hive will reveal pollen being brought in by foragers. This is followed by a frenzied bottom shaking, known as the 'Waggle Dance', which signifies the direction of the forage in relation to the sun. Usually, a series of bees will take turns to do the dance in small groups upon entering the hive. Given that the hive is extremely dark, it's the vibrations rather than the way the bees are actually facing that shows the others where to go.

I set up my hive in the back bedroom of my flat and the bees could come and go through a hosepipe that led out onto the crumbling window ledge. Below the ledge was a 50-foot drop, and severe crosswinds cut across their front porch making conditions hazardous at times. The poor bees would come out of the protected safety of the observation hive and immediately be buffeted by strong gusts.

My first mistake was to move the damn thing to the front of the house and onto the kitchen table one night before going to bed. The warm sun coming through the bedroom window had been worrying me as I thought the bees might overheat under the observation hive's glass walls. What I had foolishly failed to realise was that the flying bees were already programmed and knew where to find the entrance to their hive. To them, the ceramic white tiles of the nineteenth-century mansion block, and the scraggy buddleia hanging over the parapet, were the signs that they were almost back home after a long harvest trip.

By the morning, all the foraging bees were bumping against the back bedroom window, queuing and stacking up, wondering which idiot had moved their front porch. Moving bees even short distances, you see, is a tricky business. To the relief of its occupants, I immediately moved the hive back to its original position. I had learnt my lesson about how finely tuned a bee's navigation system can be.

Persuading people to like bees

By mid-June, my mission to make London less dependent on imported honey is finally gaining momentum. I can start moving bees out of the North London super-site as I have a growing number of satellite sites. As well as a fantastic place in East London, I have been offered a site in Hackney and another in an old building near Spitalfields Market.

Step by step, my bees are spreading across London. As usual, most of these moves have to take place in the early hours, not just to avoid the heat of the day, but to prevent the alarm of residents when they see me arriving in my white suit and veil and carrying bee boxes. These are clandestine operations organised like military raids.

I've learnt from experience that not everyone welcomes bees. Years ago, when I had hives in Greenwich, I met an elderly couple who lived nearby. They complained about my bees visiting their garden pond. The chap was allergic to them and nervous about the sight of hundreds of bees queuing up to have a drink. When I went over to make amends, I immediately understood his concern. Their pond was like bee Heathrow: there were hundreds of bees gathered close to the water, drinking, and at least another 500 waiting their turn behind.

It turned out this couple were easily convinced. When I explained my bee mission and the importance of keeping bees in the city, for pollination purposes, they quickly came on board and became tolerant towards their thirsty visitors. They were an admirable couple, like Tom and Barbara in 'The Good Life', virtually self-sufficient in food, keeping chickens and goats, catching rabbits for the pot and growing every imaginable vegetable.

That first visit, I left with an armful of fried apple rings, made from apples from their own trees. After that, I'd take them honey which they'd put in delicious home-made cakes; in return, I'd go home laden with artichokes and rabbits for my supper.

Meetings such as these make everything worth it. Even if people are initially wary of living near a hive, once they realise how integral bees are to life, they are normally quick to lend support. It also helps if you explain that bees generally won't bother people. Unlike wasps, they are not often aggressive. They are so busy on their missions, fulfilling whatever duties are relevant to the time of year, that they will not generally take any interest in passing humans.

The old pumping station

The first of my new locations – and the one I'm most excited about – is in an amazingly lush area, close to the River Thames, which I stumble upon by chance on one of my many quests for fantastic cake. I'm there at a café on a cool June Sunday, chomping down a fancy tart and reading the papers – a rare moment of relaxation. When I leave I start talking with the owner, and before I know it, she's offered me a site.

This place is now a trendy café, restaurant and gallery but it was originally an old Victorian water-pumping station. Various industrial elements of the era still remain, including eight rather splendid water tanks up on its roof. These colossal containers, now empty except for one, are going to be the homes for my hives. The one containing water has amazing lily pads that the bees end up using as landing platforms when they're thirsty.

In the summer, the art gallery is planning a series of outdoor film screenings onto a huge canvas so I'm quick to suggest showing my old bee movies from past beekeeping forays to Paris, New York and Rio, and perhaps offer some honey popcorn or ice cream for the revellers – all helping to promote the fantastic honeybee.

The site is surrounded by mature lime trees and the forage will be fantastic as it is just as green as Bermondsey. The rooftop is secure and protected from the elements but the hassle is going to be getting everything up and into the old water tanks. Not only are they five storeys from the ground, but the tanks themselves are ten metres deep; they present an immense challenge, not only for getting the hives in but for getting any potential honey crop out.

I need not have worried. With the assistance of Josh, the bold Aussie who co-owns this place, the hives are easily installed. He sees it all as a breeze. Balancing each individual hive on top of his head, like an African lady with a calabash, Josh descends into the abyss down a wobbly ladder wearing nothing but a Pacha Ibiza T-shirt, shorts and flip-flops. Within each massive tank, I help to set down the hives on their raised stands, made from old pallets, which have also been hauled up and into them by the great man.

My business is unnervingly reliant on figures like Josh. His approach, manner and dedication to whatever I throw at him are inspiring. I'm planning to organise another volunteer day later in the month, a chance to get together as many bee-friendly people as possible to help out with all the tasks that need doing. This gang of helpers will mostly be building frames and assembling hives as I just don't have enough time or energy to do it all on my own.

Moving bees when they have fresh nectar within the hive is also very hazardous. Energy is already being used by the bees to expel moisture from the nectar by the frantic fanning of their wings, and they will struggle to keep the hive cool at the same time. Even more care needs to be taken on these humid nights.

My only concern about the pumping station site is about the atmosphere inside the steel tanks. I'm particularly worried that the intensity of heat will be too great for the bees in the warmer months, and that they and their combs will suffer. I'm also concerned that the sides of the tank will keep out much of the available sunlight in the winter.

In the past, I have used green builders' netting to shade the hives and prevent wax from slipping to the bottom of each frame. I often find it in skips – a great resource for the urban beekeeper. I stretch it above the metal roofs, allowing air to circulate and sun to dapple the hives rather than roast them.

Although sunshine is generally good for bees, the temperature several storeys up on an asphalt roof with a sunny vista can be considerably more in the height of summer than at street level, especially if there is no breeze. You should position the hives carefully with evenly distributed weight as legs can sink through soft, warm bitumen. It goes without saying that you should also check that any potential rooftops can take several hundred kilos in weight. A hive falling through the rafters would be unpopular. I have heard of melted honey dripping through a ceiling cavity, and though sticky it might not be quite so bad.

Each hive on the pumping station is placed to maximise the early-morning sun. This helps the bees to flourish generally; I also find the warmth gets them out of bed and foraging early. As a rule of thumb, a south-facing beehive is essential for the best bee health.

I've always relished the chance to watch the last few bees returning to the hive at the end of the day. Here, I can stand above them on a gantry, with views of the financial City, the river and London's parks, and watch those hard workers returning after sunset. I'm beginning to think this will be a great satellite site.

Three days later, I return to see how the bees are settling in. I'm relieved to find that they are thriving and can be seen (to the trained eye) from street level spiralling out from the tanks like wisps of smoke. My hunch was right; it's a fantastic site and it's great to see the little creatures prospering. In the back of my mind lurks the thought of the huge amount of lifting required to get the honey crop up and out and down, but with Josh on hand, I try not to worry. He assures me he has great plans involving a winch newly purchased from Lidl.

Tasting different varieties

At a honey festival organised in Battersea Power Station by organic box-scheme company Abel & Cole, which is attended by over 300 beekeeping and honey devotees, I perform a special tasting. It involves four of the most delicious UK honeys, produced by David in Wales and by myself, arranged around the edge of a plate.

Imagine the plate like this. At twelve o'clock, there is borage honey, a flower grown mainly in East Anglia for its oil which is used in the pharmaceutical industry, but also put as a blossom in glasses of Pimm's to make them look pretty. In the past, I have traipsed across the country with several hundred bee colonies for this unique harvest. The bees go wild for the bright purple/blue flowers. Borage produces a light honey which is almost translucent when fresh, and then when it sets, it turns pure white. The audience love it but sadly borage production has mostly shifted to China so it's not an easy British honey to get hold of.

At three o'clock on the plate, it's Salisbury Plain honey, which has a mildly medicinal smell. It's delicious but hard to describe, and this is probably how honey used to taste in the UK, before oilseed rape was introduced and dominated the forage of most bees in the countryside.

At six o'clock, there is one of my London honeys and a vintage version too. It's the last of my Bermondsey rooftop crop from about eight years ago. It's toffee-like and is a big hit.

Finally, at nine o'clock is the strongest of all the honeys, which is why it needs to be tasted last. This is UK heather honey, rich in protein, and to my mind infinitely better than the overrated manuka honey. It is our most highly prized honey as it requires the most exhausting expedition to gather it. Weary beekeepers have to trawl around on top of the moors to gather it, watching the weather like hawks so as not to miss the brief window when the heather flowers.

Tasting honey is surprisingly similar to tasting wines. You start with a delicate flavour and finish with the strongest one so you don't destroy your palate. You take your time to breathe in the smell of the honey – never dive straight in without having a good sniff first. It's amazing how nuanced, delicate and dramatically different honey strains can be. This is why I lay them out so carefully and I advise the guests to have a glass of water after each chomp.

Summertime occupations Afterwards, I'm approached by one of the managers of Hackney Marshes, who is keen to get more bees onto their site. It's the break that I have been waiting for as it means I can have another super-site like North London, but this time to the east.

I'm going to have to move quickly, though. Peter always warned me that North London would yield the majority of its honey at the end of June. The question I face now is whether to move bees into my new Hackney site immediately or wait until after the main flow of honey in case I disrupt it. It's a huge dilemma that could be costly if I get it wrong – the sort of dilemma that all commercial producers, be they beekeepers or farmers, sometimes have to deal with.

Meanwhile, the threat of swarming is still high, though not quite on the same red alert as it was last month. Sensible beekeepers will keep up weekly checks to look for signs of swarming in their hives.

I've also got some deliveries to make, and one of these is to Fortnum & Mason. They have been my client for almost ten years now and I am privileged to be their bee-master. Where possible I try and deliver orders in person to their tiny delivery entrance; the friendly staff are a pleasure to deal with and always want to know how the bees are doing on their rooftop.

The buyers are very knowledgeable folk. Sam Rosen Nash, for the Grocery Department, has recently started keeping her own bees. She is always one of the first to race up onto the roof when there are reports of swarms appearing in the West End, to check that the bees are still there – which they usually are.

This recipe of Sam's features their honey. I adore this shortbread with strawbs and double cream – sort of mashed together à la Eton Mess – or, as Tom Bean created on the heather moors in Yorkshire once, with a dollop of thick clotted cream on top, a raspberry and finally a tiny piece of gelatinous heather comb. It's tricky to wrestle into your gob, mind you …

Summer Shortbread

....

4 tbsp F&M heather or lavender honey
225g butter
350g plain flour
15g cornflower
a pinch of salt
caster sugar and fresh strawberries, to serve

Cream the honey and butter, then fold in the flour, corn-flower and salt. Knead the dough on a floured surface for 3–4 minutes until smooth. Wrap in cling film and chill in fridge for 30 minutes.

Roll out on a floured surface to 1cm thick. Using a cutter, cut out small discs and place onto a lightly greased baking tray. Bake for 10–12 minutes at 160°C/ 325°F/gas 3 until lightly browned.

Serve sprinkled with caster sugar and fresh strawberries.

Bumblebees and wasps By now, I have started receiving calls from people who want bumblebee's nests removed from their gardens. I always question why anyone needs to remove a bumblebee's nest in the first place. I think it's partly because they think they are wasp's nests, although the two are quite easy to tell apart. Made from a papery material, wasp's nests are brown and grey and hang down, like basketballs, while a bumble-bee's nest looks more like a bird's nest; it's normally under-ground or buried in the undergrowth, and consists of tiny little pots of honey.

Wasps I can understand being disliked as they can be aggressive, but bumblebees generally just bumble around. I try to convince people that it's a privilege to have one of their nests in your garden, especially if it's in a woodpile or somewhere where it's not bothering anyone. Obviously if it's very close to the house, you might want to have it moved to a nearby wood, but since it's illegal to kill bees in the UK, you'll have to find an alternative home for it.

Fetching new bees

If you are building up your bee colonies, or getting your first bees as a beginner, this is the time of year you will be most likely to collect a nucleus – a young starter colony which will include five or six frames along with a young queen who will have mated by now and be ready to start laying.

It's another occasion on which you will need to deal with the many dangers surrounding moving bees in a vehicle. A nucleus will probably be contained in a vented box, roughly the size of two shoeboxes. It will have lots of air vents and the lid should be screwed tightly shut.

As well as making sure this box is not placed in direct sunlight during the journey, you will also need to check that it has plenty of air around it and a cooling draught – just as you would a normal hive. That most cars have air-conditioning nowadays is a help, but it is still important to plan your journey and travel later on warm summer days.

When I'm collecting nuclei, I always think it is a sign of a passionate breeder if they take the trouble to explain exactly how to take care of your bees in transit. Most responsible bee breeders will stress the importance of delivering this precious cargo safely to its new home. You are responsible for over 5,000 inhabitants within this fragile box. Some beekeepers have told me they've gone so far as to snatch their tenderly nurtured colonies from the back seat of a new purchaser's car when they have felt the bees might not be well cared for.

Other bad habits that novices might accidently pick up include wrapping young nucs in towels, old curtains or rugs to prevent leakage onto precious upholstery during transit. It's difficult to understand how anyone can expect delicate bees to survive a journey in a hot car when shrouded in fabric.

Collecting a nucleus

Things to remember when you collect your nucleus:

- Ensure the carrying box has good ventilation.

It will usually have mesh windows glued into place on the side or top, to allow a draught to flow in and out. Stopping regularly on your journey to open doors and windows and allow air through the vehicle is a good idea. Think about where the box will get the best flow of air.

- Choose your time of day carefully.

If it is a sweltering day, see if it is possible to collect your new residents in the evening, when the bees have stopped flying and when the temperature has cooled down.

- Bring a water spray, filled with fresh water after you've made sure it is completely clean and chemical-free.

Spraying bees with a little water every now and then can help cool them.

- Position the box carefully in the vehicle so that the frames face the same direction as the car.

Peter taught me this when I used to collect bees from him; it means the frames do not swing around and crush bees.

- Secure the box so it can't easily be dislodged.

I learnt this the hard way some years ago when I had a box of bees on the front passenger seat and had to brake suddenly. The piece of foam blocking the entrance shot out as the box crashed into the footwell, releasing some of the bees. I had to abandon the car until the evening when the confused creatures had settled in a cluster on the ceiling.

- Make sure you always have some gaffer tape, along with a drawstring mesh bag and a few straps.

Some people use seat belts around the box, but in a car I think it's better to wedge it down and then secure with a hive strap. Most of my vehicles are open air nowadays so any bees that leak have to cling to the outside of the hive. The mesh bag can be put over a box or hive of bees as an extra layer of containment that doesn't compromise ventilation.

- Plan your expedition.

Make a weekend of it and think carefully about everything you will need. When I sold bees from Shrewsbury, I used to have couples asking if I knew any good bee-friendly B&Bs for the return journey.

- Always bring your bee suit, just in case.

Another reason to travel late in the day when collecting a nucleus is so that you have the cooler hours to address any disasters en route. The most alarming thing to see is lots of bees leaking out of a box, particularly if they are in the car with you. It's especially worrying – and distracting – if you are driving, as they often appear darting frantically in your rear-view mirror.

To see just a few is quite normal – it could be stray bees that have clung to vents on the outside of the box, in which case you can simply stop and let them out of the car. It's pretty impossible to get them back into the box and you just have to regard these bees as lost. But if you see more than a dozen, you have a leak. It's time to pull over, don your suit – which you should always keep to hand when transporting bees – and address the situation.

The reason for a leak is usually to do with a dodgy box. It could be old and a little rickety or it might have a wobbly roof. Bees are ingenious at finding ways to escape. This is why I always carry some gaffer tape so I can patch up any problematic holes.

Once they've arrived at their final destination, I position each nuc box alongside the new hive the bees will be transferred into the following morning (with their front doors facing the same way), and I try to open them at dusk to allow a few bees to fly before darkness sets in. Don't leave them in these tiny nuc boxes for long. Generally the colonies will be extremely strong and virile and will swarm if there is no space, especially if a honey flow arrives.

It is surprising how quickly bees will adapt to their surroundings. They rapidly familiarise themselves with their new turf and set to work on a bounty of forage. When I take bees to the heather moors, I usually find that within twenty minutes of our arrival, they are arriving back into the hive with fresh pollen.

A father to my bees

Over the summer, a beginner beekeeper should take time to get acquainted with his or her new brood. There's so much fascinating stuff to learn about them but until you feel you know what you're doing this can also be an anxious time. When I started out, after a day's work I would lie awake into the night, worrying about my bees' health, before rising tired and dark eyed to begin the cycle again.

Those anxious early days eventually passed. The paternal hormones were kicking in and I was falling deeply in protective love with my charges. On warm evenings I'd sit in front of the hive with a cup of tea, waiting for the last foragers to return – like a father waiting up for teenagers to come home after a late night.

Worst places to be stung

In my years keeping bees, I have had some terrible disasters on the road. The most unpleasant for me personally was being stung on my testicle. Foolishly, I was wearing shorts and one escapee managed to shoot up my leg and sting me – oh, the pain (and the swelling later) … I managed to carry on driving and just about keep my eyes on the road, but it was an experience I wouldn't wish to repeat.

At the time, it reminded me of a story that had been in the news about Mick Jagger, having deliberately got his penis stung by bees in order to enhance its size. This ancient Amazonian marriage ritual was reportedly conducted while he was filming scenes for his 1982 movie 'Fitzcarraldo'. Enough to make your eyes water just thinking about it, but it is actually not the most painful place to be stung. That right is reserved for inside your ear and the end of your nose – especially if you suffer badly from hay fever.

Major transportation

Transporting large-scale quantities of hives in order to harvest a specific seasonal nectar is a different operation altogether, and needs even more attention as numbers are usually larger and require more space. I have experimented with a variety of vehicles over the past ten years from an old Morris Minor pick-up to an out-of-use London black taxicab. Both kept the bees separated from the driver, which is always a good thing. The cab had a glass partition (and of course an intercom for chatting to the ladies). Someone did once try to get in the back when I stopped at the lights, which would have given them a nasty shock.

In America, bees are moved in their thousands to pollinate huge groves of almonds and melons. Loaded up by forklift onto pallets and stored in giant juggernauts, the bees spend several days cooped up inside and experience huge stress as a result. They are considered nothing more than a commercial commodity, something that can easily be replaced if lost. Things have got so bad with bee disease in America that huge colonies are now imported by plane from Australia and New Zealand in cardboard tubes to replace collapsed colonies. All of which suggests that it makes sense to look after bees carefully when moving them.

Thinking locally for varied flavours

There is one big journey I have been making for some time with David, the annual migration of the bees on to Salisbury Plain at the end of June. This allows them to take advantage of amazing wild flowers, such as Viper's Bugloss, a flowering herb called sainfoin and the sweet clovers that cover this rich wilderness. It is a special time for me. Not only do I catch up with friends who live that way but the complete wilderness experience is very evocative and pleasurable. I'm an outdoor guy and this is my chance to have a fantastic adventure, catch some great wild fish and camp under the stars – weather permitting.

This year it's raining and we arrive in the middle of military manoeuvres – the sky is alive with flares and we set down the bees in virtual daylight, despite it being 1 a.m. This vast expanse of land is a place of contrasts. The plain is kept undeveloped and remains an unspoilt wilderness where wildlife prospers, especially wild flowers; yet because it is an area of great scientific interest it is heavily regulated and guarded. It is also used consistently by the British army as a training ground, and tanks and troops romp around it.

These bee adventures are not just about a boys' expedition, though; they are the chance to try and produce more unusual honeys that will offer something different to the customer and hopefully expand their palate.

David has supported me on numerous attempts to try and source more unusual honeys from around the UK. One of these was sea lavender on the east-coast marshes of Mersea. We thought our bees had gathered buckets of the delicate nectar, as the hives were positioned right amongst the purple flowers. However, when we tested the result it turned out to be white clover honey. The bees had ignored the delicate sea lavender and gone for the copious lush clover not far away.

Postponing the Hackney move

As for my London bees, I decide not to move them to Hackney just yet. Or at least, the weather makes this decision for me. By the end of June, the huge honey flow that was forecast due to good weather has not arrived. I decide to keep the bees in North London and wait for a honey crop before setting up my new East London site. Even in the capital there are some areas that can yield loads of nectar, whilst others remain barren for weeks. But I know good weather must be on its way and it would be crazy to move hives when lots of honey is about to pour in.

The nectar flow needs to arrive soon, though, or I'll be in real danger of having starving bees on my hands. I spend the tail end of June racing around them all with gallons of syrup. The month finishes with only a spit of stores in the corners of the brood chambers. It's a very worrying time because feeding bees on the point of a honey flow can trigger contamination of honey when the flow does eventually arrive. Sugar syrup can make a difference to the taste, so it's the last thing you want in your honey boxes.

What new beekeepers should be doing:

- Young nuclei of bees are usually ready for collection from bee breeders with their newly mated queens. Their new residence should be ready for their arrival with fresh frames and feed on standby.
- You will need to collect your bees and plan the journey home with them very carefully. Transfer them out of the nuc box and into the hive the next morning.
- You are no longer a potential new beekeeper but a real one. Get stuck into your new life. Joy!

More extensive beekeeping tips:

- Check the flow of bees in and out of the hive on most days as a quick guide to what is going on. You should still be doing a full brood inspection on a weekly basis.

- Look out for increased activity at midday when the young bees head out for their acclimatisation and orientation flights. It can look as if the bees are massing to swarm, but normal traffic soon resumes.
- Hive nests can become colossal and you may need to consider splitting colonies. New-season mated British queens should be available from renowned breeders and will be handy for replacing older queens or for splits.

JULY

COLOSSAL NECTAR AND HONEY FLOWS

So long as the weather doesn't play unpredictable tricks, July is a great month for collecting honey. At my North London bee site, the colossal nectar flow predicted by my friend Peter has finally arrived. It's an enormous relief. I was getting nervous about whether a honey crop would ever find its way into the hives – I should have had more faith in my old teacher.

These past few days, all my hard work has been coming to fruition, but just as in *The Sorcerer's Apprentice*, the cup has begun to overflow. I'm rapidly running out of honey boxes to house the mighty and valuable crop.

I've been so busy concentrating on my bees' welfare and feeling pessimistic about their honey-producing ability, this sudden change of fortune has caught me unawares. My bees are not just flourishing; they are revelling in the sweltering temperatures and the nectar paradise offered by the city woodland. I have barely considered how I am going to sell my precious crop, except I'm already set on the idea of keeping as much on the comb as possible.

An incredible greenish floral lime honey has begun surging into the honey boxes. I had always been taught that a strong bee colony could fill a box in a few weeks. But if the conditions are perfect, which indeed they have been recently, it turns out that bees can fill boxes in a matter of days.

As I pile more and more of my dwindling honey boxes on the hives with their winter-assembled honey frames inside, they look like illegal high-rise skyscrapers. I am even forced to use brood boxes – double the size of honey boxes – to catch the bountiful liquid. This is in some ways a poisoned chalice as each box will weigh almost 50kg when full, making them difficult to move. I can't believe the bees will manage to fill these as well.

Ordering enough hives and boxes should have taken place over the winter months. But if you are anything like me, you will end up doing it in a mad rush as and when you need things. This can mean you forget to make careful checks for quality in your equipment. It's particularly important that you use fresh new foundation wax if you're making honeycomb as you won't be extracting the honey but leaving the combs behind, keeping them as part of the package.

At this time of year bees will be having a wild time feasting on nectar supplies, using their long tongues like straws to suck it out of flowers and trees. It is then stored in their special 'honey sac', a separate stomach unique to bees. Back at the hive, worker bees use their mouths to suck the nectar from the foraging bees' honey sacs, and then chew the nectar for about half an hour – it is not 'bee spit' as described by my nephews. This breaks down the complex sugars into simple sugars so that the nectar is both more digestible for the bees and less likely to grow bacteria while it is stored within the hive. Next they transfer it into the comb and as a final flourish, when the moisture has been removed, the bees seal off each cell of the honeycomb with a plug of wax.

Extra frames in a hurry

You always need to make sure there's enough space on the frames in your honey boxes. Like me, in a pinch you can give them brand-new frames at this stage if you really need to, but it's far from ideal when the nectar is flowing. They will have to take time out from processing nectar in order to draw wax out into cells before they can be filled.

There are special bees dedicated to this job of building combs. You can see them hanging down off the frames in a line, clutching onto each other. Wax platelets are secreted from the bottom half of their abdomen and passed up the chain to the bees at the top; these then use their ingenious lower jaws, called mandibles, to mould the wax into the hexagonal shape.

If you are planning to extract your honey, then a better option than providing brand-new combs is to use what are called 'drawn-out' combs. These are combs the bees have built in previous years and therefore ready for them to fill with nectar. They can immediately get on with packing their cargoes of delight into them.

To make a new frame, you need to insert a piece of foundation, which is a beeswax sheet held together with wire, into the wooden frame surround. It encourages the bees to build combs in a structured manner. You can make this up yourself – the cheap option that us professionals go for – but suppliers will also offer these sheets with the wire already in there. You just need to tack them into place with another bit of kit, the unfortunately named gimp pins. Again, you don't want to be faced with doing this when the honey is flowing. You want lots of frames and combs ready to go so you don't hold the bees back.

The first year I kept bees, I had no idea that you could use frames year after year, or indeed that it would be beneficial to do so because your bees wouldn't have to work on new ones. Foolishly, I threw away my first honey frames. I now have some that are twenty years old and still in good working order. It's a slightly different story with brood frames, which contain the combs where the young brood is reared. They become very dark after a year or two and reusing them is said to encourage disease, so I tend to invest in new ones at least every other year or carefully sterilise and recyle old ones.

Frames or brace combs Failing to provide new honey boxes altogether, however, would be an even bigger mistake than making the bees work on new frames. After every conceivable space had been crammed, you would end up with what is called brace comb (also known as wild or burr comb), a big pancake of comb that bees create if their production is not controlled with frames.

Brace combs can look very impressive if you don't use frames or see pictures of them in countries like Zambia where they are produced as a matter of course. They hang like huge dinner plates if there's room in the cavity, with just enough space between to allow a bee to pass through and a tiny bit more. We call this the bee space. Frames are spaced in a certain manner, leaving just enough space to prevent this brace comb being built but big enough for the bees to move through.

It's worth remembering that brace combs are how most of the developing world still deals with its honey, and it's a more natural system that has become a popular way of keeping bees recently; harvesting the honey is a real challenge, however, as is getting the bees to move off the combs. It is not a system that would work well on a hot rooftop, when you need the manipulation of your bees to be as straightforward as possible.

Over a hundred years ago, bees were still kept in straw skeps, small domes of woven straw that allowed them to build these wild combs. Because no internal structures were provided for them, the bees created honeycomb that was dense and could not be moved without wrecking the hive and often killing the bees – clearly not a very sustainable method. Honey would be extracted by crushing the wax honeycomb to squeeze out the honey – a method which still exists around the world and which I have witnessed throughout Africa and Asia.

Our modern hives are more practical and effective. You can pull out individual combs in the frame with little or no damage to the colony, making manipulations easy and controlled.

Lots of lovely honey

Except this year, I'm struggling to come up with enough honey boxes. I've heard of beekeepers putting cardboard boxes on the hives to collect wild comb, but this doesn't appeal. Instead I phone a manufacturer late one evening and plead with him to fire up his workshop. As ever, because it's the middle of the season, the honey boxes I eventually secure turn out to be expensive – almost double the winter price – but faced with this emergency I have no choice.

In the past, huge manufacturers of bee equipment had been forcing out smaller hive-building craftsmen, but this is gradually changing. A plus side of the resurgence in beekeeping as a hobby is that a new breed of younger, skilled woodworkers has sprung up, offering affordable, good-quality products. Some are ex-builders, forced by the recession to turn their hand to something new; all offer an alternative to the giant makers, who have had a stranglehold on the market for too long.

For most beekeepers, removing a honey crop when it becomes ripe will be one of the main features of the month. When nectar is brought into the hive, it starts off very thin; as much as 80 per cent water. The ripening process involves the evaporation of moisture to thicken the nectar; made faster by the bees fanning their wings inside the hive. If you reach an apiary in the evening and hear this fabulous noise, thousands of bees using their wings to fan, it's a sure sign that they have been busy working that day.

Fresh nectar sometimes drips onto your feet. If a frame is tipped horizontally, nectar can pour out like water, a sign that it has not ripened or been sealed with wax. You need to be cautious about wasting this preciously harvested crop. You'll know when your honey has ripened because it will have thickened, assuming the consistency of runny honey.

Honey is one of the only food substances that can never spoil. It can change its characteristics as the natural sugars crystallise, but it won't go off. So long, that is, as the beekeeper doesn't remove the honey before the bees have sealed each cell with wax. In this case, it would ferment very quickly, and customers would not appreciate bubbling yeasty honey.

Ripe honey is thick and sealed over with a tiny white capping by the bees; it is now ready for removal. I know some beekeepers like to harvest different crops across the year, which means you get to taste some really unusual and special honeys, especially in London. I used to do this – but now I guess I still do the same with combs, which represent a particular variety of honey.

At this point in the season, I could remove my crop but as yet I have nowhere to store it – I don't seem to have time for studio-hunting – so it just sits on the hives. That's why I continue to pile boxes on top as each becomes full.

Experience has taught me that, luckily enough, London honey seems to maintain perfect texture for several months whereas honey from other parts of the UK can crystallise. Fortunately too, theft has never been a problem for me – the burglars would have to be beekeepers – though sadly I hear it is on the increase as both honey and bees are worth a tidy sum. I have even heard of full boxes of honey being stolen from hives, their honey being extracted and the empty boxes returned for refilling.

The downside of leaving boxes on the hives until next month is that it will mean stripping several very heavy boxes off the top of a hive in the full heat of the day, before even reaching the bottom brood box to start a hive inspection. This is completely masochistic, but despite the fact that bees are less likely to swarm now, I still need to check hives thoroughly every few weeks.

My failure to provide adequate storage room for full honey boxes slows down the visits I can make in one day as I spend my time heaving off the honey boxes I've had to leave in situ. It always amazes me how one simple mistake can escalate quickly in this industry and become severe almost overnight. I've got to find a base.

If you decide you do want to remove your honey boxes and start extracting honey or cutting up the comb, you'll first have to get the bees off the frames you want to pilfer – a delicate process which I describe in the section 'Removing the bees' in August (p. 177). My preferred method involves a leaf blower, but most people starting out will probably stick to a technique that involves something called a bee escape, or clearer board, a one-way valve that allows bees to leave the honey but doesn't let them back in.

Water for bees

Bees need water. It is essential for maintaining a constant temperature within the hive and aiding cooling when necessary, as well as for the building of wax and diluting crystallised honey. For a colony to prosper, it is essential that you make sure they have access to a regular supply and if need be provide some. If you keep bees on a rooftop, which is often a warm suntrap anyway, it's even more important to make sure they always have a drink handy.

More than a fresh running tap, bees seem to love dirty, stagnant water. In the countryside you'll notice them gathering around murky puddles, paddling around in the soft mud; they can take on minerals and use the mud as a landing station.

Preferring not to piss in a pot for my bees, instead I set up an old upturned dustbin lid for containing water, with a few bricks for them to land on. The water will slowly become grubby and stagnant and that's what they love.

It's also generally believed that in hot weather, your honey crop will be greater if there is water on hand for bees. If you don't provide it, then they will seek it elsewhere and this can cause problems, with large numbers of bees queuing up at other people's small garden ponds, dripping taps and puddles.

Marauding wasps

At this time of year, there's also the problem of wasps to consider. Back in early spring, when I do my first checks on my hives, I always discover chubby queen wasps slumped underneath the roofs of my beehives. They enjoy its warmth and dryness when they are hibernating. I'm sorry to say they are quickly evicted. Come July hundreds of wasp offspring will otherwise be trying to overpower my hives and intimidate my bees.

Hives that are weak or young are at risk of being swamped by wasps in the summer months. Smaller colonies have insufficient numbers of occupants to offer any proper defence and will just watch as the wasps steal their honey. You will need to take extra care to avoid this. A strong colony is able to repel these intruders with force – the bees will often dart out of the hive at speed and knock them to the ground.

I opened a hive on Salisbury Plain last year that had more wasps inside it than bees. The queen bee was still there and a few workers were clinging to the frames, bullied into submission by the marauding invaders – perhaps weakened by swarming, what had once been a prospering hive with several boxes of honey was now looking at a bleak future.

Apart from doing your best to build up strong bee colonies, your only other option is to reduce the physical size of the hive entrance, making it into a fortress that can be more easily defended. On remote bee sites, this can be done by simply shoving tufts of grass into the entrance, and restricting the passage to just the size of a ten-pence piece. Some hives come with specific wasp guards – or you can buy them separately – to prevent your precious crop being pilfered.

One question that is often passed my way is about the purpose of wasps. To most of us, they are evil critters that terrorise beer gardens, yellow-and-black tormentors who transform adults picnicking on a warm day into hysterical loons who wave their arms around frantically.

In fact, they are highly intelligent pollinators that I am told have even been used by the military to detect explosives; they are also carnivores who usefully chomp on dead bees dumped outside the entrances of hives, or, to the delight of gardeners, on live greenfly.

This year seems a particularly bad year for wasps and I notice them more than usual on the urban sites, especially the more central ones. Perhaps they are attracted to the West End by all the sticky ice creams and sweet cans bought by tourists; the worst affected is a new site on the National Portrait Gallery off Trafalgar Square. I need not worry, however, as the staff there are alert to these pesky raiders and they have made crude traps – each a plastic bottle of water with the top cut off and inverted into the body, then topped up with jam and cola. Carnage.

When I display my bees at markets and shows in London, in an observation hive, I am amazed by how many people think they are wasps. I often get calls from people saying they have bees in their house, which then turn out to be wasps.

That so many people get them confused surprises me as wasps are very distinctive. They are clearly yellow and black, slim, and smooth where bees are hairy or downy. A wasp's sting is not barbed like a bee's and will continue to sting you if the beast becomes trapped in your clothing. They are generally aggressive, compared to bees which are divine – but they have their purpose in this world as pollinators and despite being evil they are highly intelligent creatures.

Wasps are certainly always able to detect ice cream in the summer months in the capital. Fortnum's Jonathan Miller, the chap who designed the glorious beehives on their rooftop, bought us all ice creams on the hot summer's afternoon when the bees were installed. Within minutes we were troubled by yellow and black invaders – wanting some of the champagne float which I had created.

I worked on an interesting recipe for ice cream with Olive magazine. Their version used honeycomb brittle, actually made from caster sugar and bicarbonate of soda, but which for some strange reason people often confuse with bee-produced wax honeycomb. It worked well but I suggest you use some fresh honeycomb, which gives it a little interesting texture. Try and find some fresh local British summer runny honeycomb that is floral and clear – even better if it has some fresh pollen in it.

Double Honey Ice Cream

225g runny honey
600ml double cream
250ml full-fat milk
6 large egg yolks
honeycomb, strong or aromatic – I use London or
*   Salisbury combs*

Warm the runny honey in a small pan over a very low heat. In another pan, preferably a double boiler, heat the cream and milk together until just below boiling point.

Whisk the egg yolks together in a large bowl until combined.

Pour the hot cream mixture over the eggs, whisking constantly. Return the whole lot to the double boiler and set over a low heat, stirring constantly, until the custard thickens enough to coat the back of a wooden spoon.

Strain into a clean bowl, add the warmed honey and mix together thoroughly. Leave to cool and then chill for at least two hours before churning in an ice-cream machine following the manufacturer's instructions. (I don't have a machine so I whip the mixture in a haphazard way with a fork every now and then as it freezes.)

You can either add the comb broken into chunks when the mixture is cool or towards the end of churning so as not to completely break up all those lush chunks. I add extra at the end for a total honey overload....

Oilseed rape When the good weather breaks, I decide to return briefly to Shropshire, partly to help David prepare his bees for the heather harvest next month, and partly to pick up some more equipment that I have stored in my old cowshed.

My visit reminds me of one reason why I'm thrilled to leave the countryside behind. Oilseed rape, an insipid mono-floral crop, is everywhere at this time of year. Although it is revered by some beekeepers since the bees can produce a bumper harvest from it, the honey produced is poor quality. It is clear, smells of cabbage and lacks depth – not very appealing.

Worse than that, however, oilseed rape can do considerable damage to bees, as they inevitably become over-stimulated by it. It's a powerful bee magnet, and they are somehow captivated by it while ignoring neighbouring fields of bluebells and clover that make delicious honey. They soon become exhausted, their wings ragged and their tiny systems over-hyped, and yet they still can't resist going back for more.

Oilseed rape is bee crack – and just as damaging. In addition, they are at great risk from being killed by the insecticides and pesticides with which it's incessantly sprayed.

Some of the worst times of my beekeeping career have been on intensively farmed vales near fields of this glorified cabbage; I have found huge numbers of dead and decaying bees in rotten flesh-smelling piles outside the entrances to their hives. Such a gruesome discovery would be distressing for anyone, let alone a bee-loving man. The bees are not just my livestock and livelihood – I hate to see them suffering. Over the past decade, I've been forced to close five sites that were too close to a crop of rape.

It is now grown with real intensity across large parts of the UK, and can spring up as early as April and last until late September. It grows terrifyingly quickly: with the introduction of numerous fertilisers and the right weather conditions, it can flower in less than a few days.

When I lived in the countryside, I would always scout around an area to check there was no rape nearby before depositing my bees. Frequently I would return just a few weeks later to discover the hives encircled by the stuff.

Another problem is that oilseed rape can be genetically modified, occasionally causing previously unknown viruses to come to a head. It's no longer a natural flower source; instead it's engineered in a lab, and once it's fed to young bees and grubs, it can do untold damage. One of the most traumatic viruses for a beekeeper to witness is called Chronic Paralysis. You lift the lid on your hives and see bees shivering and quaking with this condition, a sight which never fails to bring tears to my eyes; and while its cause cannot be solely attributed to oilseed rape – the weather is another factor – I have witnessed bees struggling in intensely farmed areas.

In London, there is nothing like this to put my bees at risk, although I did once find a single plant of oilseed rape in my friend Doug's garden in Blackheath. It was crawling with bees. No idea how it got there. I immediately marched over and asked Doug for a bin bag and pulled out the root. It's the last thing I want to see flowering and prospering here in London. His girlfriend was a little alarmed at my guerrilla approach until I explained.

My stay in Shropshire only lasts a few days, but it's long enough to remember that country life, instead of producing nectar nirvana as I had once thought – with fresh, unpolluted pastures – in fact posed one of the biggest threats to bees. I'm glad to be away from all the fields of yellow. Besides, there's so much to do in London; I can't leave my hives for long.

Mandana A few days after my return a potential disaster strikes, but fortunately this time it's me who gets hurt and not the bees. I attempt to lift a hive, forgetting about a brick that I have positioned on the roof. There is a dull thud as the brick hits my nose, sending blood shooting out into my veil and blocking my vision. Proud of the fact that I didn't drop the hive, I quickly click my nose back into place; I've broken it before so I know the drill. Witnessing the drama, Mandana, one of my most diligent volunteers, lets out an initial yelp when she turns to face me, but she still manages to complete her meticulous hive check. She is not the pampering sort and we crack on with inspections – I like her no-nonsense approach and she is a great bee apprentice.

To keep up with honey production, I find myself working later into the night, using huge battery-powered spotlights to illuminate the hives while I add more honey boxes. Although some people have found the opposite, I find I get stung more at night than during the day. The bees can't see so well, and they don't have the same instinct to fly away, so they cling to me and work their way inside my clothes.

I know of a newish beekeeper in Battersea who has taken to inspecting his bees at night because he is not confident with them. How pleasurable must it be to have bees that you can't handle during the day? Weird. It sounds as though he needs to re-queen the hive and get some gentler ladies in.

Mandana is like a diligent terrier, forever at my side, battling away at what often seems like a losing scrap. She is a huge help, never asking for payment, always turning up with her crisply pressed white bee smock and her sparkling new brass smoker, which is often admired but never used. Having got badly stung a few weeks ago, she now wears chef's white trousers over her own and then wraps her trouser belt around the outside of her smock. It's an effective technique popular with beekeepers – and surprisingly stylish.

She is slim and elegant, but despite once turning up with a Vivienne Westwood clutch bag, she always appears to be up for a challenge. I have never seen her baulk at a task. Some beekeepers are reluctant to let others into their hives, but I feel confident in Mandana's ability. She is slow but diligent and I know she will not squash Her Madge, the queen bee, or aggravate the other bees into a frenzy. She is a gentle soul.

Her family fled Iran when she was young and she was schooled in London. She is heavily involved in Soho life and runs an old-fashioned writer's club, where she plans to install a few beehives to produce honey for the restaurant attached. I love her devotion and the way she approaches things. She knows that she wants to learn this craft properly and that it will take perseverance, so she offers up her precious time freely and learns from me in the process.

David's London Some people just want to blast straight in and keep bees immediately. I was asked last year if you could learn everything you needed to about beekeeping from the Internet – obviously not. Everyone wants knowledge instantly today, but beekeeping is not that sort of activity; it's an evolving process.

No one understands this better than David. In the late 1970s, he used to have twenty-odd hives in London, scattered across three sites in Clapham, Brixton and Surrey Quays. Like me, he had aspirations to try and achieve city bee dominancy, which I guess is why he is so supportive of my project.

There are other similarities with the way he used to operate his bee business, supplying small delis and stores, and my own attempts to do the same. He wanted to bring a bit of the countryside to the city and see if it could prosper – which of course it did. Like me in my early years of honey production, he could never produce enough to supply demand, which I am hoping will change with this year's bumper crop.

Far more than me, he was a true bee pioneer, pottering about between his London hives, his own Morris Minor van never far away. Original wholefood pioneers Neal's Yard, based in Covent Garden, bought everything he produced and recycled the jars which customers would bring back.

In Brixton, on Josephine Avenue, the site of his first hives, bees were positioned in the rafters of a squat and would spiral through a hole in the roof where the tiles were missing. I once asked him if his Brixton bees had ever produced marijuana honey, which I had heard could be achieved thanks to the large quantity of cannabis plants growing in local gardens.

David dismissed this. He used to import a Jamaican logwood honey and the beekeeper assured him that bees would avoid the pungent sticky buds produced by marijuana plants.

He also added that you should never use ganja in your smoker to pacify your bees as it would have the opposite effect, making them crazy with rage. This spoiled my image of sleepy bees lounging around on their backs, eating loads of honey and becoming a little bit paranoid.

David is familiar with many of the problems I've faced this year, such as weighty honey boxes after a failure to anticipate the amount of honey that would come in. This is why I often turn to him for advice. He has his own tales of lowering honey boxes by rope from the rafters of the Brixton squat when the intensity of the nectar flow produced some back-breaking crops.

At David's Surrey Quays site – back then, a derelict area still awaiting the yuppie revolution – there was a purple sea of buddleia that would produce exceptional honey. But David talks with mixed feelings about another kind of honey that came from honeydew, a sugar-rich sticky substance that makes a thick brown, almost blackish honey.

Honeydew Honeydew honey is highly prized by some; not least because the way that bees collect it is rather unusual. It does not come from nectar produced by flowers to entice bees to pollinate them, but instead it is a secretion that emerges on the backs of insects – in the UK, this is usually an aphid. This sugary liquid, produced in huge quantities, is collected by both bees and ants. It's also what causes those sticky smears on your car windows in the summer which are always impossible to remove.

The flow of fruity-tasting honeydew begins in July and can help bolster a poor honey crop – especially if conditions are humid, which will encourage a plague of aphids. It is more apparent in some areas of London than others and it is believed to be an indicator of the poor health of trees and plants. In short, it signals an infestation of aphids and possible plant diseases.

Personally, I love the stuff. Like a truffle, it is unique and not easy to find, but that makes it all the more exciting when you see the dark frames appearing in your hive. I try to separate it out from the other honeycombs for special treatment as its production has an element of alchemy. What fascinates me is that honey can come from such an unusual substance. Most people assume that without flowers or other trees and plants, bees couldn't make honey. When in fact there exists a dark art of honey production from an insect secretion.

Smelling and tasting a little fungal – OK, so I know I'm not selling it very well – honeydew gives the bees a chance to gather a honey from something different, when there is little else available to them. That strikes me as ingenious. A downside is that in large quantities, it can be rather toxic to the bees; it is high in yeast and the bees become dizzy and dozy on it, almost pissed.

Bermondsey is a good area for honeydew production and I'm hoping that the same will be true of the new sites on the two Tate galleries, not so far away.

The Tate gallery sites From the second week of July, as soon as my Tate pass is issued, Mandana and I begin installing hives. It's exciting stuff and we pick a cool evening to move a dozen hives down from North London to their new luxury apiary.

We focus on setting up six hives on the Tate Modern first. The move will be one of the hardest this year as the site where the hives will be positioned is some distance from the lift shaft, which means a lot of heaving. Luckily, the gallery has been really well organised, arranging porters and trolleys for us, and Mandana's long arms fit perfectly under the hives in order to lift them from the truck onto the trolleys in the loading bay. Each must weigh about 35kg but moving them is more about technique than brawn. In my opinion women make the best bee farmers. They are calm with the bees, but also tough and dig in hard when the going gets tough.

The gallery has even installed lockers on the stairwell near the hives, crammed with crisp white beekeeping clobber which puts mine to shame. The gear is not only for me, but also so that visitors can meet the bees.

When we arrive at the staff car park and press the buzzer, late one afternoon, I'm sure the security officers will be wary of letting us in, but they seem well briefed and the barrier shoots up. The bees sound quiet and content in the back of the van, a good sign since they are about to be manhandled by porters. These turn out to be cleaners from Brazil, who, far from being clumsy or insensitive, are a huge help, barely worrying the bees at all.

Each of the half-dozen hives, painted bright yellow, will have one of Tarpaulin Mike's mesh bags over the top, in case there are any leaks – or worse still, a hive is dropped. This has happened before in a lift while I was wearing shorts and resulted in more stings than I would like to remember. Tate Modern is a labyrinth of chambers, lifts and corridors and it takes some time to even get close to the roof. Any weeping bees will be hard to retrieve as the ceilings tower above us in this former power station, so the mesh bags are to stop me worrying. The last part of the journey involves carrying the bees up four long flights of stairs and then finally out onto the roof.

Swinging from my T-shirt is my new Tate pass, the thought of which keeps me going as I climb the stairs yet again with another load. One of the best things is that it entitles me to free teas as well as access to some great, affordable food in their staff canteen – a good incentive for a flagging beekeeper. I insist on carrying the first few hives from the lift to their new spot, but I'm puffed pretty quickly and end up accepting help from the porters, who manage one between them.

At twelve storeys up, this apiary space will be my highest this summer and the views will be stunning. My hives are being kept in a well – a sort of sunken area on the roof. By evening, its walls are glowing with the fabulous hue of the setting sun, and the light is bouncing off the tall glass building north of the river. It's magical stuff but in the back of my mind is the worry that if they swarm, their likely resting point is the thatched roof of the nearby Globe Theatre.

Down below, the air is thick and humid, but up here, it's fresh and cool on our sweat-clad backs. The breeze subdues the bees and their intense fanning gradually dies down. Mandana and I position the hives on their wooden stands and we move the entrance blocks a little to allow the bees a small flight before the light finally fades. A few gently leave, which is a good sign.

This wouldn't be advisable in the middle of the day because they'd all shoot out and start trying to make sense of their new location. But as the sun sets, it's cooler, the bees are calmer and only a handful emerge from the hives to investigate. They don't really have the inclination to go far away from the hive at this time of day; that'll be left until tomorrow when they can start exploring nearby forage.

When all of the security staff have been thanked and bid goodnight, and Mandana has left for home, I put into action my secret plan. I sneak back onto the roof and unroll my mattress. I'm spending the night here as I'm worried about the direction that the bees will fly in the morning, and whether they will prosper at this height. This kind of overnight stay should not be encouraged, but in my defence, it's in the name of my bees' welfare.

As I settle down for the night, I'm reminded of an organisation that celebrates covert sleepovers in famous buildings around the world – Alcatraz Prison and San Francisco's Golden Gate Bridge, to name a few.

While the night is magical, I sleep badly. Even up here, London is noisy with sirens and traffic and then at dawn, the aeroplanes start roaring across the sky. As I dose lightly, around half five, the sound of the bees beginning their navigational flight wakes me. The buzzing noise is immense. They appear to be heading south towards the railway lines past London Bridge Station. The scrubland is fantastic forage for bees as it remains relatively undisturbed and wild – and long may it remain so. Reassured that they're settling in OK and are seemingly undisturbed by the dizzying height of their hives, I creep off to find coffee in the staff canteen and start my day a little grubby but happy that all is well at altitude.

By the end of the month, I've also installed six hives at sister museum Tate Britain. It's an easier job as the roof is only two storeys up, meaning the site is much more visible and overlooked by several offices; my activities provide great entertainment for all the office workers who crowd at the windows to watch.

Swarm in East London Just as I'm beginning to relax, satisfied that the Tate hive installations have gone well and that the risk of swarming is waning, I get a call from my Aussie volunteer Josh. The intensity of the month's honey flow proves to be getting my other bees excited and a swarm has dropped down from the water tanks on the old pumping station. He tells me it is surrounding an ornate piece of brickwork above the kitchen in the restaurant underneath.

Apart from the manager who is panicking about a wedding party due the following lunchtime, everyone else is very cool about this when I arrive. Josh is as keen to help as ever, sorting a ladder and a cardboard box at my request, but I already know this is going to be a tough one to remove as I am concerned the queen might have climbed in through an airbrick.

If she has, she'll be impossible to remove. It will be fatal for her, not to mention a disaster for me as I'll lose a queen and have great trouble getting the other bees away from her. There's not a chance I can dismantle a listed Victorian building so I hope that she is still within the cluster.

The bees are calm, but when my nose comes to within about six inches of the swarm, a little shimmer goes across them, a sign that they know I am there and are on high alert. They can sense any tiny movement and certainly the carbon dioxide from my breath. When you see a collective shudder across a swarm of bees, it's a reminder to be on your guard as it shows that they are feeling threatened and could do something unpredictable.

At last I find the queen, wedged behind some brickwork. I gently pick her up with my fingers, making sure not to squeeze her abdomen. A queen's tummy is very delicate; hold her the wrong way or accidently squeeze her and you can ruin her reproductive system. An old goose feather is stuck in my back pocket and I use it to gently brush the rest of the bees into the cardboard box. Disaster averted.

Zambia I'm lucky to have support from my bee friends and mentors. David is confident my London adventure will succeed and this sees me through the tough times. He is a modest bee farmer and I respect his opinion.

Much as I'd like to tell you that the stereotypical features of a beekeeper, such as a scraggly beard and mad scientist eyebrows, are in fact all a myth, it would not be the truth. All beekeepers can be eccentric showmen at times and David is no exception. Many people have picked up on his dark and foreboding presence. He is a man of few words, except of course if you get him talking about bees. But he is an extremely straight, honest and knowledgeable chap. Without his support, my grand bee project would be floundering.

Although he is fifteen years my senior, I'm constantly impressed by his stamina and fitness. He works like a Trojan when he's moving and hauling equipment, never holding back when it comes to mucking in. Years ago, we used to employ young students in busy bee months and even they would struggle with the workload.

David and I met over twelve years ago, when I was still a photographer. I had managed to persuade the Body Shop to employ me and my then girlfriend to document the Zambian honey harvest and I'd heard that David was an expert on Zambian bees and hoped he'd come with us.

In 1984, he had first visited north-west Zambia as part of a VSO (Voluntary Service Overseas) placement. There he established a community of beekeepers, who formed a cooperative harvesting honey in the traditional manner with techniques exchanged over hundreds of years. Over time, he has grown his business in Wales so he now imports Fairtrade organic honey and beeswax from over 6,000 beekeepers across the world, often in places where families have no other means of supporting themselves.

One of the problems he has faced is the tendency for NGOs to encourage local people in developing countries to use Western-style beehives. These hives sit like abandoned coffins in yards, as they are neither practical, nor popular, in such communities.

The Western style of beekeeping on frames is not workable in other parts of the world, where bees have been kept in traditional hives for hundreds of years. Here, people make hives out of whatever is around them. In Cameroon, West Africa, it is woven grass; in Zambia, it is tree bark. If a beekeeper needs money, he crops honey from his hive and sells it, much like going to a cashpoint and making a withdrawal.

For our Zambia visit, we took bags bursting with climbing ropes and harnesses. It was definitely a case of all the gear, no idea. We had gone to extreme measures with equipment because we had heard about the way that beekeepers keep their bees in this part of Africa.

I had seen faded snaps of bark hives and they were mostly up in the tree canopies – about 90 feet up. We knew that photographing them from the forest floor would have been pointless. We wanted to be among the action, even though these bees were renowned for a terrible temperament. These were not domesticated bees, they were Africa's finest and wildest.

Even more worrying was the fact that my girlfriend was badly allergic to bee stings. Her face would swell to frightening proportions with just one single sting. At first it was funny – she looked like the elephant women – but over time it was alarming and I would receive shocked looks as passers-by tried to work out if I was responsible for her bruised, swollen face. But she was a fantastic photographer who loved bees and nothing would stop her getting involved.

Once we were in Zambia, the equipment proved to be a mixed blessing. The best pictures were taken by tying my girl-friend to a rope and hoisting her up a tree with the help of all the villagers. Due to our inexperience at using harnesses, I still managed to fall 15 feet out of a tree. I wasn't badly hurt, except for my pride. The entire village witnessed my fall, which must have seemed even more amusing as we had all this unnecessary equipment. Local beekeepers climb up to the hives without a single rope or harness.

The temperament of the bees was nothing like as bad as we'd heard, but we did hear terrifying stories about another aggressive animal. One of the reasons Zambians keep their hives so far up trees is to keep them safe from hungry honey badgers that will destroy the entire hive to get at the honey. These sweet-toothed mammals, black in colour with grey and white backs, have a fearsome reputation. One elderly villager assured me that a honey badger could kill a bull elephant just by leaping up and biting his testicles. We were relieved not to encounter any during our visit.

As for the bees, their stings were no worse than those of UK bees, but we took no chances and wore full beekeeping gear at all times. When we went home, we left the veils, gloves and bee outfits behind in the village. I wasn't sure if the local beekeepers would bother wearing any of the kit, but they seemed very grateful at the time. Who knows, maybe they're just using it as fancy dress.

Back home, the trip was counted a huge success. We had some great photos for the Body Shop, and they asked me to install four hives at their main office in Littlehampton. This was to be part of a bigger picture there, with allotments where staff could tinker in their lunch breaks, and solar panelling across the roof, giving them a chance to learn about the environment in a hands-on way. Bees were the obvious next step. Now owned by a giant French cosmetic company, the Body Shop remains true to its founder's roots and the bees formed a natural link to their philosophy.

Urban beekeeping of a different type as the site was close to downland as well as towns. I installed some fine Welsh Bees. More importantly on a personal level, the trip also garnered good publicity for David's company, and at the same time was the beginning of a smashing relationship for me.

What all beekeepers should be doing:

- Make sure your bees always have access to plenty of water throughout the warm weather.
- You will still need to check your bees weekly for signs of swarming but the risk will be greatly reduced when the colony peaks this month, about three weeks after the longest day.
- Look out for wasps that can suddenly appear and torment your bees. I put jam traps out to capture them and reduce the entrance size of weaker hives to prevent them from being overrun.
- A massive nectar flow can begin if the conditions are right so ensure you have enough honey boxes ready for the rush.

- Order jars and labels in anticipation of your huge honey crop! (Unless conditions have been incredibly good, this is unlikely to apply to new beekeepers.)

AUGUST

EXTRACTING THE HONEY

Yorkshire heather It's the start of August and I'm helping David with his colossal bee manoeuvres. We are relocating dozens of his hives to Yorkshire's bleak moors on the east coast. Centuries of beekeepers have transported their bees to this part of Yorkshire to take advantage of the stubbly beds of flowering heather, particularly a variety known as ling, that produces the finest and the most highly priced honey in the UK.

As always, the challenge is in the timing. To get the best and the most honey, we must release the bees just as the heather is budding and bursting into flower. They will bring in the first nectar within a few hours of arriving at their summer holiday destination, which is always a huge thrill.

The honey boxes need to be close at hand so they can be attached as soon as the bees have familiarised themselves with the mass of purple. But our truck will be overloaded with hives so we resort to using a courier company to bring up the boxes a few days before. They drop pallets of them on the sides of remote arteries across moorland roads; the boxes stand out like blue Tardises, ready to be mopped up by our teams once the bees are in position.

Logistically, the coordination requires immense planning and we call on a few experienced hands. These volunteers are more than happy to help out, keen to be involved in this ancient beekeeping adventure.

Many years ago, specific trains were laid on for bees to be shuttled to the moors of Yorkshire, so common was the expedition among beekeepers. Today, I can't see Virgin Trains laying on special coaches for the job. Besides, I think David and I both love the adventure and the thrill of the harvest chase, and we enjoy doing as much as possible ourselves.

Heather is a weather-sensitive crop and the ancient peaty ground requires prolonged drenching for weeks beforehand to ensure there is enough moisture for a drawn-out flowering period. Then, in an ideal world, the wet period should be followed by sticky humid days, without strong winds, to provide perfect conditions.

At this time of year, there is nothing on the remote moors for the bees to forage on except heather. It's a complete contrast to bee forage in London, where it would be impossible to provide one pure form of nectar, although the lime harvest does come close.

Because of this, the resulting honey is the closest you'll get to organic honey, something that is officially unachievable in the UK since you can't guarantee where 50,000 bees from each hive will travel to. (It's worth remembering if you find organic honey in the shops; read the label carefully as you'll probably find it's been imported.) These isolated moors, with only one flower in bloom, create the nearest thing.

Traditionally, beekeepers believe they must have their bees on the heather by the Glorious Twelfth, the 12th of August, which is start of the grouse season. Vast expanses of the Yorkshire moors today are still managed commercially for their grouse. This works well for bee operations like David's and mine as the heather is cut short to encourage the green shoots that the young grouse feed on. When the heather blooms, it is young and vigorous with lots of new growth. So long, that is, as there isn't an infestation of heather beetle, which can quickly reduce the sea of purple to a reddish rust.

A young gamekeeper who I come across, rifle slung over his back and several spaniels perching diligently on the rear of his quad-bike, tells me that a gun on the moor this year would set you back £10,000 and that is with a limit of one brace (two grouse). A costly hobby.

When we're positioning David's hives, we try to get them as close to the heather as possible, if not actually sitting on it, to prevent the bees using up energy by flying. Before we leave them to their holiday, we remove the queen excluders. These have been keeping the queen in the brood box and preventing her from laying in the honey boxes, but honey production is thought to be more efficient without them. They slow the bees down as they have to work their way through the mesh opening to get into the honey boxes. By now, since the queen will be laying less, she is not so likely to move up into the boxes; and even if she does, any brood will have hatched and their place with any luck been filled with honey by the time the boxes are removed in six weeks' time.

If the weather behaves itself, the Yorkshire trip can reap glorious rewards, but just to be sure, we embark on a second stage of our heather expedition, which involves taking hives to the opposite side of the country and the heather moors in North Wales. By doing this, we hedge our bets. There's always a slim chance that both sides of the country will have poor weather, but it is unlikely.

On our way to Wales, David and I share driving and rest periods. I'm already on first-name terms with the cashier at the A1 service station, just north of Leeds. These truck stops are a welcome oasis despite their poor-quality coffee. It's an opportunity to stock up on nibbles. My theory is that it's impossible to fall asleep at the wheel if you are eating and this year my preferred snacks are Bombay mix, Mars bars and thin smoked German sausages. David, on the other hand, always chooses pistachios; he drops the shells into the footwell, which can make clutch control a little hazardous as they pile up.

Neanderthal man

Having helped distribute David's hives around the country, I now need to ready some of my own London hives to ensure they catch a heather harvest. For this, I'm taking about a quarter of my total number of urban hives to the hills in Shropshire, and I have a tipper truck for this job as I think it will take a substantial load. I don't technically need the tipping facility but it might just come in handy for unloading.

This mission has already taken its toll on other areas of my life. I've neglected my current squeeze, a headmistress, who is sitting on a beach in Crete, waiting for me to arrive. The actress turned out to want more glitz and glamour, sweetie, and not a crusty unwashed beekeeper with wild hair and cages of queen bees in his pockets, so I was binned. [Cycling back over Waterloo Bridge months later, I see her name up in lights on the side of the National Theatre – she's made it big – and despite being a little pissed on my Boris bike, I'm delighted for her.] I expect my not turning up in Crete will mean yet another relationship biting the dust.

The nomadic life doesn't do a bee farmer any favours on new dates either – after even only a day in the field, getting ready can be an art form. I often arrive at the location of the date straight from a moorland or other outpost with only a few minutes to spare, which means concealing the rustic look as quickly as possible with the minimal amount of effort. I have learnt that it is not very well received in London by the opposite sex.

Bee smocks and overalls are removed with haste. Walking boots or wellies must never be worn – I'd do better to go barefoot. It's all about forward planning and I usually have a fresh set of clean clothes and jacket – pretty much shrink-wrapped to guard against smears of honey – on board my truck.

Disguising the fact that I could be a twenty-a-day man is a little harder; a week in the field wrestling tricky bees with copious amounts of smoke means that the smell is deeply embedded in my skin. Added to the fact that I may not have washed for a week which means I hum a little. Bingo ... toxic wino look.

Sometimes desperate times call for desperate measures and I have been known to spruce up with a car air freshener – those lemon Christmas trees aren't that bad really – but nowadays I try and keep a deodorant somewhere in the truck. If time permits I will wash in a service station basin – not a full strip wash but at least face, hands and hair. Nails are a little trickier as they get filled with wax, propolis and general grime – so to mask my Neanderthal tendencies I keep my hands in my pockets for as much of the evening as possible.

Shropshire heather Whether or not it is either lucrative or practical, the Shropshire heather-harvest expedition has become a pilgrimage for me. I do it because I have done it for the past twelve years. It has become something of an obsession; a homage to the moor and the amazing honey that it can produce and, of course, to the ingenious, resourceful honeybee.

As kids growing up in Shropshire, we used to play in the hills during the school holidays, damming the streams that contained wild trout and sucking the rich perfumed stimulant from the heather flowers on tacky August days. We thought it was pure honey.

It is still the highlight of my beekeeping year, especially when everything comes together to create a good heather harvest. There is something magical about arriving on a moor in the middle of the night and unloading your bees, then returning a few weeks later to see what they have achieved. It's a drug. I could never give it up – the allure is too great. I do it because I love it.

But I can see that it is also a little whimsical. It makes little financial sense to move bees from one place, where honey could still be available, to somewhere else, especially when the heather harvest is, at best, a gamble. Fuel is expensive and for the past few years the bees have not made enough heather honey to cover the cost of their transportation. Each pot of liquid gold would have to be sold at £50 just to cover costs. No one would buy it for that, so the move is heavily subsidised by my London bee operation.

Despite these drawbacks, this year, just like every other, I'm on my way. I set about finding my most productive hives to take. There is no point moving bees that are young and still establishing themselves. I need hard-working grafters.

Then there's the question of trying to condense an entire hiveful of bees into a brood box. First, you need to get all the bees out of the honey boxes since these will be removed. There's no point bringing London honey to Shropshire so I only transport empty ones to collect the heather honey. Once the boxes are taken off, it is like trying to pour a tin of baked bees into a thimble. Huge mounds of bees hang from the entrances, unsure if there's space enough inside, when I return to load them up.

My usual tactic is to use my smoker and gently encourage them into the hive. It's like getting commuters into a crowded Tube train. You never think they will all fit, but it's amazing what you can achieve with a bit of nudging and encouraging. Some clamber underneath and work their way up inside through the mesh floor, something you have to watch for when you're lifting a hive; the last thing you want is a sting on the hand.

I leave London with my elite set of buzzers packed carefully on to my hired truck. As usual, to help me with this year's Shropshire haul, I've invited my friend Kingy, the gamekeeper whose sticks I have been using in my smoker; he manages 5,000 pheasant on a Shropshire estate. Since I was a boy, I've been fascinated by gamekeepers. I love to learn from these country almanacs, who are generally resourceful, hardy people, and who were practising green living long before it became fashionable.

Back during the Second World War, gamekeepers trained elite forces in the Highlands in survival techniques and how to live off the land. Kingy, who is in his sixties, is a fantastic companion and the trip has become a social event as much as anything else. We make sure we're accompanied by a small bottle of sloe gin and a rough picnic. This usually includes a slab of cheese, a bit of pickle, some bread, and always a tin of mackerel, packed into an old wicker basket.

Kingy always brings his own bees on this trip. His beehives are made from timber scavenged by Barry the Ferreter, a one-armed retired builder who is a demon at his craft. Each hive is unique, except for the roofs, which are all covered in aluminium from a scrapped caravan, carefully shaped around the edges to provide a waterproof seal.

Kingy's hive numbers have swelled to over twenty colonies now, most of the bees having been procured from hedgerows and hollow trees. As soon as he hears about a swarm, he straps a swarm box to the back of his bike, hangs a pair of binoculars off the handlebars and sets off in hot pursuit across the fields.

Once bees have swarmed – and particularly if their beekeeper fails to pick them up promptly – taking ownership of the bees is a grey area. Whose are they? It's very hard to reclaim bees once they've been collected by someone else and especially if they are settled in another hive. Over the years, Kingy has boosted his numbers by being quick off the mark and most beekeepers agree that this is an acceptable way to procure bees.

Kingy's bees are some of the strongest, best-cared-for stocks around. Each hive is given a brand-new queen, reared using his incubator – the sort used for hen's eggs, but which Kingy actually uses for rearing pheasant. Inside the incubator, the queen cells are placed in his wife Julie's hair rollers to keep the young queens safe when they emerge. Otherwise, they might either wander off or more likely, they might have a battle if more than one emerged at the same time. This could result in damaged queens.

Since a queen bee needs to feed on honey within half an hour of emerging from a cell, sometimes the best solution is to put the hair roller inside a hive so the other bees can feed her as soon as possible.

Kingy is diligently waiting for me when I arrive on the moor at 2 a.m. He's not asleep in his car but instead his smoker is lit and he's pacing the track; I can tell something is worrying him. 'There's a caravan parked where we usually put the bees,' he says. This is a huge problem and we set off into the paddock to inspect the ground and see who has taken our prime location.

Hesitant about disturbing any sleepers, we approach through the trees with caution, only to discover it's a compost toilet erected for the Duke of Edinburgh Award hikers who will be passing through next month. It's a huge relief but even so we can't put down hives near this temporary bog; visitors might well be distressed by the huge numbers of bees out foraging and coming and going from the hives.

We quickly search for a new location, something which I hate doing at night as you are never quite certain what hazards are where. We settle, after what seems like hours of deliberating, on the opposite corner of the paddock. I'm hoping the firs will not shadow the hives in the mornings and that the footpath is far enough away from the flight path of the bees – it's a gamble, but I'm knackered. When we are done, Kingy offers me the spare bed at his cottage for a few hours' rest – luxury.

I stir late, but as always enjoy the views from his windows of the hills and the coppice on the Long Mynd. It's pouring down – good luck, Bees. I never leave Kingy's cottage empty-handed. Bantam eggs, rat bait, ex-army jumpers, gas masks, even antihistamines for my hay fever are some of his recent gifts. He tried to give me a racing pigeon last year. He was forced to give up this hobby after developing pigeon lung, an allergic reaction caused by their droppings, feathers and dust. It causes shortness of breath, coughing and feverish illness, so marking the end of this craze.

My bees will spend about six weeks in Shropshire. I might visit them once or twice towards the end of the month to see how they are doing, but I won't collect them until October.

Shrewsbury Show There is time before my return to London to check the honey displays in the Shropshire Beekeeper's Honey Show Competition, which forms part of the annual Shrewsbury Flower Show. I have visited this since I was a small child.

Several years ago, honey exhibits were thin on the ground, with the majority of entrants coming from more mature beekeepers. But this year, the show's 124th, I'm delighted to see a fantastic array of honey delights throughout the various classes – including honey biscuits, made to an age-old recipe, simple but fantastic, which junior entrants have to follow every year. It was my first memory of honey exhibiting as a child and I was very proud a few years ago to see my nephew George's biscuits with a second-class rosette on them. Some were a little caught on the edges but I knew they'd still be delicious as he'd made them with the previous year's heather honey.

Class 72 Honey Biscuits

Junior Class
Group 1 for Juniors up to 11 years of age

1st Prize £6.00
2nd Prize £4.00
3rd Price £3.00

1 tbsp honey [George my nephew thought a strong honey
was best]
50g butter
½ tsp bicarbonate of soda
50g plain flour
80g rolled porridge oats
50g granulated sugar

Put the honey and butter in a basin and melt in a microwave oven. Take the basin from the microwave and add the bicarbonate of soda. Blend the flour, oats and sugar in another bowl and then mix them with the honey/butter mixture and allow to cool.

Divide the mixture into twelve walnut-sized balls and place on a greased baking tray (not too close together). Flatten them slightly. Bake at gas mark 160°C/325°F/gas 3, for about 15 minutes. The finished appearance will look more like cookies than conventional biscuits. Please exhibit 6 biscuits from the 12 produced.

Young people attempting these recipes should be supervised by a responsible person at all times.

The only thing to add after 'exhibit 6 biscuits from the 12 produced' is that you should give the other six to your uncle for his trip back to London....

Taking off the honey boxes

Back in London, I'm hoping the bees I left behind will still produce a final batch of honey, but I can't fault them as they have already delivered a great harvest this year.

I've been slow to remove honey from my urban hives because of a problem with where to store it, and I'm not worried about it crystallising. But for most beekeepers, August is the month to remove your summer crop of honey when sealed and ripe.

Most people, having ordered their jars some time in July, will now be thinking about the process of extracting honey, the sticky art of removing your prized crop from the honeycomb cells. I choose to leave it until next month, not just because of a lack of premises, but also because I'm pretty stretched with transporting my bees across the country in the hope of catching some flowering heather.

Since few beekeepers go to the great lengths that I do to move bees, then apart from extraction, August should for most not be a stressful month. The bees are pretty stable by now and starting to slow down, and as swarming is also unlikely, it might be a good time to take a well-earned holiday.

It's worth mentioning that you may find your bees a little more aggressive towards the end of this month. As the warm weather and foraging season begin to draw to a close, they know it's their last chance to make honey; they become defensive about protecting supplies. After all, a colony's survival over the winter depends on whether they have enough stores. This becomes even more the case next month, when the weather cools, but it is still worth paying extra attention when you check on a hive, if only to avoid stings.

When you come to remove your honey boxes, do remember that it is a delicate job and it must be done carefully and respectfully. At the same time it needs to be like a military exercise; honey must be removed efficiently and with minimal fuss, so as not to give the bees any hint that you are about to pilfer their valuable crop. If they get wind of what's going on, they'll be more likely to rush into the honey boxes and eat as much of it as possible, which will mean less delicious crop for you.

The trick is to start encouraging them away from the honey boxes without them realising what you're up to. I have tried doing it at night, so as not to disturb them, but you need to ensure all your bees have left their stores first and this operation is harder to oversee in the dark.

Bringing indoors honey boxes that are still full of bees, particularly in tiny flats in London, should be avoided at all cost. Should you do this, plenty of the bees will become excited and fly towards windows and lights. Trust me, I've inadvertently made this mistake and not only is it a bit terrifying having so many bees in your flat, but you will also risk the chance of your honey extraction area becoming contaminated with suicidal bees who, intent on a quick fix, dive into buckets of extracted honey.

Removing the bees There are several ways to make sure your honey boxes are free of bees before removing them. When I am in a hurry and the weather is warm enough, I use an industrial leaf blower, which straps onto my back. This 2-stroke petrol beast features a cone-shaped nozzle which I direct down each seam of bees still residing in the honey store.

To do this, I position the honey box on its end, on top of a neighbouring hive, and blow the bees quickly and firmly from the combs and into their returning flight path. By adjusting the hose length and nozzle size, you can direct the bees so that they end up near the hive entrance, to stop them from becoming too disorientated. The blast needs to be sufficient to remove them, but gentle enough not to damage them.

It takes a fair bit of practice and I would not recommend this method to a beekeeping novice, especially in an urban environment as you'll always end up with lots of bees in the air, which can be alarming. Blowing bees on rooftops can also be hazardous since they are often inadvertently directed into air-conditioning vents or cooling fans.

The bees appear unaffected by the sharp gust, and after a few minutes will congregate back at their allotted hive entrance. If the weather is too cool, however, they will gather for warmth on the blower operator, which is why this is only an option in warm weather. In the past, doing this in cool weather resulted in huge clumps of bees seeking shelter on the back of my head. Failing to notice them until I was back in my car was a messy and painful mistake.

The great advantage of the leaf-blower method is that you can remove bees there and then from honey boxes. You only need to visit the hives once and that is a wondrous thing if you have to move bees on the same day as removing a crop.

For the urban beekeeper starting out, with less time pressure and no intention of moving bees, I would recommend an item called a bee escape, or sometimes a clearer board. These oblong-shaped, wafer-thin bits of plastic have a diamond-shaped piece of wire that tapers to an opening slightly smaller than a bee. Essentially, it allows bees to squeeze through a gap to leave the honey box, but not to come back in again. It's a crude design but remarkably effective, and cheap as chips.

Make sure, however, that your bee escape doesn't become blocked. Sometimes you'll find a fat drone bee wedged in it, its chubby abdomen spoiling the migration south for everyone else.

If the weather is poor, remember, it can take several days for bees to descend through the bee escape, as they will not be so tempted outdoors to start foraging. I sometimes put an empty box below the bee escape to provide extra space as an incentive.

You also need to make sure you have no eggs or young bees in the honey store, as bees will never abandon brood and will cling to the combs for fear that it will become chilled. This might happen if the queen has managed to squeeze through a crack or hole in the queen excluder and stroll on up to the honey boxes.

On numerous occasions, I have found a queen wandering in the stores. In which case, I gently encourage her back to the brood box with a feather or hive tool – I try to avoid picking up my queens as their internal organs are so sensitive.

There are a few other options for encouraging bees to leave their honey alone, including an oily chemical spray that sends them fleeing from the combs. I've always felt that this could not be good for honey purity, especially if you are selling it in the comb, so it doesn't appeal to me.

One final thing you need to check for when attaching a bee escape is any holes or gaps in your honey boxes. These often appear at the ends, when the boxes become old and worn by heavy use of your hive tool. If your boxes are not sealed it is possible that other bees will come in and rob the harvest, taking advantage of the unguarded treasure. Once your own bees have descended through the bee escape, there will be no buzzing soldiers left to defend the crop. Returning several days later to looted combs is soul destroying, and can be catastrophic if you lose a huge harvest. All that hard work and devotion will have been for nothing.

In the past, I've used plasticine to plug any crevices in my honey boxes, but I soon found out that bees will nibble away at this temporary plug. Now I bind them with gaffer tape, creating a tight bee-proof barrier. Then I seal up the roofs and crown boards, which sit on top of the honey boxes, as foreign bees will also try and gain access through here.

Naked honey extraction

Now that you've removed your honey, extracting it is the next job. This is a way of getting the honey out of the combs by using a centrifugal spinner – a sort of giant salad spinner for honeycombs. You put the uncapped combs in and it spins them so that the honey shoots to the outside of the drum and trickles down to the bottom, where there is a tap so that it can be drawn off.

Honey removal comes with all sorts of problems and hassles and it's important you are well prepared and have the relevant equipment. Buckets are a good example – the last thing you want is to be filling all your saucepans and pots with your precious bounty. Make sure you also have sufficient space for the job – and no gaps in your floorboards where honey might drip down. I think several kilos of valuable honey still sit under the floor of my old flat in Bermondsey, from when a bucket overflowed unnoticed.

Extractors come in various sizes and they are so costly that mine is on loan to Kingy at the moment, giving me another good excuse to put off this job. As I haven't yet got a fixed abode in any case, I don't really have any option. I'm lucky that London honey is incredibly stable. It stays runny in the honey boxes rather than crystallising and becoming brittle, which is one reason why I'm able to leave this for another few weeks.

Most beekeepers will probably choose to do it while the weather is reliably warm. My friend Malcolm's father used to keep bees with him on the beach at Dungeness, in Kent, where they produced exceptional honey from the wood sage that grows there. He always said that if you're not sweating when you're extracting honey, then it's not warm enough for the job. He used to extract his honey in a converted corrugated-tin chicken shed, in the middle of summer, wearing just his vest and shorts.

The point is that on a warm evening, the honey will flow more easily from the frames, making the job much quicker. I extracted my first-ever London honey in my kitchen, using a hand-cranked spinner borrowed from the local beekeeping association. It was so humid, and winding the handle was such hard work and made me so hot, that I ended up in the nude.

You slice off the caps, the delicate plugs of wax that keep the honey in each cell, with a very sharp knife – again, a task that is easier when the wax is warm and malleable. I use a knife given to me by an old farmer; I hate to think of its past life but it's regularly disinfected and is excellent for the task as the blade is long, firm and sharp. For those with a bigger budget, you can invest in a heated electric knife that slices through the fine white wax cappings. I have even seen some beekeepers melt these wax seals with a paint-stripping gun but this requires some skill – they pop because there is a tiny air pocket behind each cell.

Extracting in the nude would probably be frowned upon by environmental health today, especially given the furore some years ago about small producers making things to sell in their own kitchens. Tony Blair tried to put an end to this admirable creativity, a move which was fought against ferociously by the Women's Institute and eventually repealed. You must, however, register with your local council if you are selling food items from your own kitchen.

Extracting honey is a sticky job and I must admit it is one of my least favourite. The only reason I can think of for my intense dislike of this task is that it reminds me of Welsh beach holidays as a child, when my mother would cover me in sun cream and then leave me to become peppered in sand. Extracting honey brings the trauma all back. No matter how clean and careful I try to be, I always end up with honey everywhere and lots of things sticking to my elbows, just as they did on those beach trips.

I'd recommend renting an extractor from a local beekeeping association for your first few years. Buying an extractor is a costly investment and not necessary unless you're producing honey commercially. Even a basic hand-cranked extractor is fine, although it takes a lot more work than an electric version. When I was starting out, I used to call on all our neighbours in the block of flats to lend a hand – it can be exhausting.

The stainless-steel baskets within the extractor will need to be loaded carefully with similar-sized combs in order to balance the beast and prevent it from coming alive. An extractor lumbering across the floor is an alarming sight, and can easily free itself by wrenching the electric plug from the socket.

When this has happened before, I have flung myself at the shaking machine and braced it by administering a colossal bear hug. This is inadvisable and should especially not be considered if you are naked, as hair and honey do stick....

Once the liquid honey has been extracted from the combs and left to settle for a while in the drum, you can use the tap at the bottom to draw it off. An incredibly satisfying moment. It then needs to go into buckets to be stored or a steel or plastic settling tank ready to be bottled.

Sugar-shake testing for Varroa

As we have seen, checking for Varroa at all stages of the year is crucial, but August is when the mites are likely to be at their most concentrated, so it's a good time for a full test. Boffins have developed the icing-sugar shake to monitor how many mites are on a percentage of your bee stock.

First you need to identify where your queen is in the hive and keep her separate and safe. Then scoop up a representative sample of bees from the brood comb, ready to be put into a screw-top jar. The bees are best knocked first into a dry, clean, plastic bucket, as they find it hard going to clamber out on a shiny surface. With a sharp swift knock the bees will group at the bottom of the bucket and can be emptied into the jar – we use a funnel for ease. Before screwing on the lid of the jar, which should be covered in a fine mesh, add a tablespoonful of icing sugar.

The bees are rather barbarically then shaken, to cover them in the powdered sugar; they will appear like ghosts at the end of the ordeal, dizzy but recoverable. All this sounds rather harsh on the sample bees but it is worth mentioning that this is for the greater good of not just your colony but also others – regular monitoring for Varroa is key to knowing whether you have a potentially lethal infestation.

The loose sugar is shaken back out for analysis through the mesh lid onto a white piece of card covered in Vaseline. Sinister armour-plated mites are easy to spot – each the size of a pinhead, they are brownish in colour and on closer inspection you can see the blighters' legs and jaws – evil.

The count is relative and can be scaled up accordingly to gauge the level of infestation. Roughly 300 bees make for about 100ml in terms of a gauge for this monitoring. The number of mites translates as follows:

- 1–5 mites found is a minor infestation and you can treat it by scattering icing sugar over the frames through a sieve. This will encourage the bees to be more hygienic and clean themselves up, thus knocking the mites from their bodies and down through the mesh floor.
- 6–15 mites, which equates to several hundred mites in the brood chamber, is a medium infestation. It should be treatable by oxalic acid in January.
- 15+ mites is a potential problem and the bees might not make it through the winter. At this point you may just have to wait and let nature take its course.

Some beekeepers, particularly if they have a serious mite infestation, give their hives a chemical treatment in August. Plastic strips impregnated with a pyrethroid, a chemical pesticide also used by farmers, are used. The strips hang down between frames so that the bees have to pass over them for a certain period of time. These strips would usually be removed during the final inspections in October.

Mites subjected to this aggressive treatment would once have dropped off the bees and died, but it is now a pretty useless solution as they have developed a resistance to most of the chemicals. This is partly due to its overuse; plus it has been poorly administered, and of course the residues from crop spraying have also counterbalanced the effectiveness.

In the past few years, as a result, most beekeepers have moved away from chemical treatments in favour of a less aggressive approach. There are numerous treatments available on the market today – some temperature based, some organic – that you should consider instead of the harsher chemicals for these late summer treatments if your mite count is bad. I use an organic treatment throughout the year, which is sticky and contains essential oils; it can be administered during a honey flow and will not contaminate the honey. Like the icing sugar mentioned above, it encourages the bees to actively prune and clean themselves, dislodging the mites; it also changes the PH levels in the hives which the mites dislike and they loosen their grip. But my main blitz on Varroa is with oxalic acid in January.

Ditching the drones

By this point in the year, hive inspections can be reduced to every two weeks. The queen will be slowing her egg production and the worker bees will be starting to jettison male bees, or drones, from the hive. Keep an eye on your hive entrances in August and you will see workers dragging them out. It's a spectacular sight. Workers will use their jaws – or mandibles – literally to drag the drones out and dump them some distance away. Other times they're simply prevented from re-entering after they've been out flying during the day. Guard bees will stand watch over the hive entrance, like bodyguards refusing entry at a club.

The drones are surplus to requirement now and if allowed to stay would only eat precious stores over the winter period. That's it for them; their job of mating with virgin queens is done, and after a life that averages ninety days long, they die.

On London rooftops, the eviction of the drones can be sinister to witness. Rotund drone bees can be seen staggering across the roof as if drunk, desperate to get back in the hive. In the countryside you hardly notice it, as the drones become lost in the grass, but on asphalt roofs you can see huge waves of them stumbling around until they drop down dead.

I keep a small soft broom at the sites for sweeping up their carcases, which will otherwise clog the storm drains. I used to wander barefoot around on the rooftop at my first urban bee site, in Bermondsey. At first I worried about being stung on the feet at this time of year, until my bee knowledge expanded and I discovered that drones were stingless.

In general, throughout the year you will notice more bee carcases blowing around on flat rooftop apiaries than near grassy hives at ground level. The smooth surfaces of the roofs really reveal the number of bees that die off across the season. It means that you will be able to notice if worrying deposits of bees start to build up at random times of the year in front of the hives, indicating disease or – less likely in urban areas – damage from insecticides.

The lifecycle of drones is fascinating and I have learnt even more about them recently thanks to a research student from Bangor University, who spent a few weeks in David's remote Welsh valley studying their behaviour. He was trying to find drone collection or congregation points. These are unique, similar to bus shelters of the bee world. Drones hang out in these areas on warm sunny days, waiting for young virgin queens to emerge on mating flights. They often rest on leaves and hole up under foliage if the weather is bad, only to emerge again when it is fairer.

The research student used a queen pheromone strip attached to a weather balloon to collect them – he tied this to the back of his bike, which he pedalled around the lanes and fields until drones started grouping around it in large numbers. They were then despatched to Bangor University to check on their fertility, as well as the purity of the bloodline and their genetics.

I would love to try this technique in London, to discover if such collection points exist in an urban environment. There is no reason why not in the sites of North London and Hackney, but on the roof of Fortnum's I think they would be restricted by the architecture. In Wales, the points were often half a mile from the hives and in London, this could be difficult. There isn't the same open space so they would have to navigate buildings to find good spots to wait. I love the thought of drones cruising in the West End, waiting for young virgins to emerge.

Jars for Fortnum's The only honey I crop this month is from my hives on Fortnum and Mason's as the store is keen to get its hands on the stuff as soon as possible. They buy it all as extracted honey, rather than combs. I have noble plans for this unique honey. In the past, there has always been a huge demand for it, and this year is no exception, with a waiting list already set up by the grocery-buying department. With all its various press coverage it has developed a cult following, much like a vintage wine produced in a tiny vineyard.

Fortnum's was advanced in its commitment to the rooftop bee project. It was the first time such a venture had been attempted in London, and I'm incredibly proud of the honey, as I'm sure you will be with yours. On top of each jar, there is an individual tasting label in my spidery handwriting, which says where the bees have been foraging and what flavours might be picked up on. I'm fascinated by its provenance. Last year's was more delicate, but this year, although I am unsure of the exact source, the taste is citrusy – like grapefruit. This shows how varied London honey can be from year to year.

There is something very special about selling your first jar of honey – and I get a flutter in my stomach even now when I'm delivering to a new client. I am so proud of what my bees and I produce.

I do have a golden rule though, to help me save my pennies, and that is to never buy anything from the shops which stock your honey. Believe me, this is tricky as each has its own special way of enticing customers in; with great smells, buns and cakes at eye level, and fantastic window displays. It's hard to resist, but I know that once I've started, it could be a slippery slope to spending all the money I am receiving from my honey orders.

Another rule, which makes the first easier to follow, is that I never deliver my honey on an empty stomach, especially to the café and deli Fernandez & Wells that creates Soho's cheekiest and most delicious morsels.

This year, Fortnum's bumper crop has created some media interest. When I arrive to harvest my honey, I'm greeted by several television crews at reception. One is from CNN, the other a local news crew, keen to witness this rather unique harvest. My attire is often a source of embarrassment to me, and this time, it has not gone unnoticed by observant eyes within the building.

Today my overalls look as if they did not receive the hot wash I gave them last night. I try to hide their once-white greyness with an apron and patch the holes with silver gaffer tape. For shop visits I try to remember to replace my baseball cap with a New York peaked tweed cap and stick on a knotted cravat, but this is more about practicality than style as I find a hanky around my neck keeps any stings from getting through.

Jonathan Miller has spotted my scruffiness. He is keen to march me off to Jermyn Street, round the corner, which has some old-fashioned milliners and suit shops. He wants me to be dressed accordingly, in tweed plus-fours and a waistcoat, topped off with a straw boater. It hasn't happened yet, but I know he's got my card marked, and I can't help thinking that once he has his way, I'll look strikingly similar to my grand-mother in the picture taken of her in the 1930s.

The style would not just be for photo opportunities, but so that I look the part whenever I check on my Fortnum's bees. Given that I'm quite slack about wearing protective clothing, this get-up would leave me no more exposed to stings than my usual overalls, so I'd be perfectly happy to dress as well as play the part of Victorian bee-master, which I love really.

When I come to remove the Fortnum's honey – a few days after placing bee escapes below the honey boxes, so the bees can move downstairs to the brood boxes – it is less than I had anticipated. Honey yields are always hard to evaluate; however much you heft boxes and think you have a sure crop, it often appears less by the time you have extracted it. Also, poor weather over the past few weeks means that the bees have already started chomping on the crop, instead of being out flying and looking for final sources of nectar. Gaps have appeared in the middle of the combs, where once fresh honey was stored.

How much honey to expect

Keeping each of your hives firing at full capacity is the true test of a beekeeper's ability. But however much experience and skill you have, not to mention good fortune with the weather, hives sometimes just don't want to make more honey and seem to remain tiny all summer. You can perform all your different tricks and manipulations and nothing will work.

Remember, also, that it is rare for a young queen and her bees to make you any honey for the table in the first year of her reign. Her colony will not be at full strength until the following spring. So long as it is nurtured well, then that summer should see a good crop.

The other thing you may be surprised by, come harvest time, is the varied amount of honey produced by each hive – the colours can be quite wonderful. Make sure you sample all the different types; I like the light delicate ones that have a gentle aftertaste.

Despite each queen being the same age and strain and your hives being in exactly the same location, you can still find that each one will produce completely different amounts of honey. This is baffling, but can be explained by the fact that each hive responds differently to the conditions that it faces. It can also have to do with the genetics of each queen, and I would always consider the queen of a fine honey-producing hive as future breeding stock.

Piccadilly porches

My Fortnum's hives are the perfect example. Each one has had a queen that is the same age and comes from exactly the same strain of bee. The only difference is that each of the fancy hives has a different style of porch. The first, called the Mogul hive due to its Indian front door, was slow to get started but has managed to fill two boxes. Hive number two, with its Gothic-style porch, has done well, having produced three full boxes, while still working on filling up its fourth.

Hive number three, bearing a Chinese porch, has done best, filling an incredible six boxes, and while I'm not sure if it's the queen who is responsible or whether it's just been a fluke, I'm determined to use her to breed from as she's clearly done something right. Finally, the Roman hive, number four, has managed virtually nothing. It has been sluggish all year, despite being carefully groomed and tended, just like all the others.

When it's time to take all these honey boxes away for extraction, a willing team of staff come up to help me. Today, most of them are unfamiliar with the hives, but all are interested in the sticky boxes being traipsed through the store. For me – particularly my back – it's an enormous relief to have willing helpers to carry my hefty produce. My next challenge? To extract the honey and make sure it is beautifully packaged for the end-of-year rush. It looks rather industrial at the moment wrapped in black bin liners and parcel tape.

Undercover at the London Honey Festival

At the end of the month I covertly attend the first-ever London Honey Festival – at the South Bank. I am delighted to see so many beekeepers selling their newly bottled honey and also the various prices. With me I have a list of all the London boroughs because I am keen to see how they are represented.

I call it a day at eleven locations as my bag is already so heavy – but I am encouraged to see a map of London in the array of different-sized jars, shapes and labels. The stallholders have been telling me the benefits and virtues of London honey – how marvellous it tastes and why it is so special. It feels slightly naughty to go incognito but sometimes it's good to be the anonymous observer. My plans are short lived, however, as I get busted on my way out by a dear old beekeeping friend – who insists on giving me a freshly harvested dark jar of tar-like honey. It's from Hackney so that makes twelve boroughs, but I'm now convinced the total must be greater as the movement is clearly thriving. Brilliant.

What all beekeepers should be doing:

- Hive inspections can be reduced to once every fortnight at this stage.
- Your hives should be given the sugar-shake test for Varroa mites and treated as necessary either now or later. If using Varroa strips, consult the manufacturers' recommended dosage period.
- Honey boxes should be removed from the hives with your precious harvest and carefully stored in a cool place.
- If you're planning to put your honey in jars then you need to extract it during the warm weather. Local associations usually offer a hire system for extractors but you need to get in the queue early.
- If you're selling your honey, make sure you have designed your labels and that they comply with all the legal requirements set out by Trading Standards – a list of which is on their website.

SEPTEMBER

THE SEASON DRAWS TO A CLOSE

Paradise in Crete All I have ever really dreamt of is a tumbledown cottage in Crete with a bit of dusty land to grow vegetables and have a few white goats, a bread oven and a couple of brightly coloured beehives. Here, I would sit in the warm Mediterranean sunshine, enjoying rich, thick, aromatic Greek honey, which I would ladle on to bowls of sharp yogurt and fruit.

On miserably damp days, or when I need cheering up because my bees aren't thriving, my mind wanders to this simplistic fantasy. I pop open a rusty tin of honey, which is over ten years old, and I'm back there in an instant, eating roast rabbit with rosemary and chips, and lighting firework bees, a whizzing explosive that you can buy over the counter. When lit, they whirl out of control, like manic bees – highly dangerous, not to mention childish, but great fun. They should never be exported back to the UK in hand luggage. Right?

Crete is an ideal place to keep bees; it's very laid-back and the bees produce a strong but delicious thyme honey that is dark and the consistency of molasses. For many years, I visited every summer and when times are tough, I toy with the possibility of upping sticks and buying a tiny cottage over there. The brutal reality is that I don't even own a house in the UK. I've never quite had that chunk of money to get a deposit together, or if I have, I've put it back into my bee operation or into the creatures' salvation. But at last I've found somewhere to rent as my new base.

The end of the season For now, as the reality of everything I have to do this month sinks in, the warm Mediterranean sun will remain a distant pipe dream. A priority is to harvest some of my honey, either by extracting it and pouring the final product into jars, ready for the Christmas rush, or by freeze-storing the honeycombs. They will be wrapped in cardboard, sealed in a bag to prevent condensation and then frozen.

This year, I will have produced more honeycomb than ever before. I get a better financial return on it and I've noticed it's increasing in popularity. After years of educating customers on its virtues and giving tips on how to consume it, I find that fewer now ask if it's OK to eat the wax.

As well as harvesting, it is time to think about preparing hives for the winter period. Since they'll be closed up for several months, beekeepers need to do whatever they can at this stage to encourage them to stay healthy and disease free.

Although September is generally a less stressful time than the summer months – honey production is slowing down and the nest is reducing in size – I find it brings with it a degree of sadness. Summer is over and the season is drawing to a close. Even if it's an Indian summer and your bees continue to enjoy good weather, there's nothing more you can do to increase honey production. The year is winding up.

Fortnum & Mason's ornate bespoke hives in Leigh Manor's garden, overlooking the Stiperstones, during their tour of the UK prior to being installed on the store's Piccadilly rooftop in 2008.

Hive inspections on the rooftop of Fortnum's are always a joy, full of calmness away from the turmoil of people and traffic below. Watching bees returning in the late evening across the rooftops is enthralling.

Bees returning to the hive having been shaken
with icing sugar to monitor them for Varroa.

Peter's queens
arriving in the
post in tiny
cages with their
attendants and
some protective
wrapping – in
this case a flyer
from Lidl.

Ned the urban warrior.

The hoist at the old studio off Tower Bridge Road – winching in market equipment from the street below.

Having run out of normal-sized honey boxes for my Tate Modern bees I was forced to use deep brood boxes. Each weighed 45kg – it was a big mistake.

Mandana examining bees at Tate Modern on one of our weekly checks – they seem to love this sun trap at the back of the building.

At the old pumping station, Josh winches deep boxes full of honey down from the Victorian water tanks in a baker's tray.

Making nucs in the old water tanks.

The glorious world of beekeeping.

On my way to service bees at Fortnum's with my equipment in an old picnic hamper.

Open studio at the warehouse in Bermondsey – an old ironing board salvaged from a local skip is used to display freshly cut honeycomb, ready for the Christmas rush.

Luckily, to keep me from succumbing to seasonal blues, there are a few natural wonders to appreciate. The buddleia is still flowering. I love the way it clings to gutters and craggy buttresses and how it occupies waste ground, quickly covering it in purple. I've heard it was the first flower to take hold on bombsites during the Blitz. It is the heather of the south and even this late in the year, the bees can bring in a crop of honey from it.

Its sweet-smelling drooping purple cone-shaped flowers have taken over a derelict petrol station here in Bermondsey. Despite the concrete forecourt, it has become a jungle of 10-foot trees, covered in bees and butterflies that are addicted by its mesmerising nectar. Buddleia flourishes in this city and seems to settle anywhere.

Bermondsey has changed considerably in the past ten years and the gentrification has spread rapidly south from the river. In the process, one of the major casualties – from a beekeeper's perspective – has been mature trees. There is nothing worse than waking to the sound of a full-grown tree being bludgeoned by the council; I have been known to launch a deluge of abuse at unsympathetic tree surgeons.

London is talked about as a 'green' city mainly because of the abundance of trees. The trees are not only the lungs of this city; they are also hugely important for city bees as their main source of nectar.

When trees are secreting nectar, you can stand underneath one and feel the entire thing shaking with intense bee activity. Some nectar, such as from lime and chestnut trees, is toxic – oddly enough – to bumblebees (though it's great for honeybees), but that doesn't seem to encourage them to stay away. Instead they are often drawn to the very tree that leads to their doom – observant eyes will spot their tiny carcases lying underneath the tree's canopy.

Combining weaker colonies

At this time of year, if you discover a hive struggling it is a little late to think about introducing a new queen; but to save the bees from becoming lost and worthless, you could unite this weaker colony with another. We often do this on the heather when we have similar colonies that are not making the grade.

Combining two hives can give you a stronger colony, not only more capable of bringing in some honey but also more likely to make it through the winter. But you must do it carefully or the bees will fight and kill each other.

The way to do it is to take the floor off one hive and sit it on top of the other brood box, having first laid down a broadsheet of newspaper in between. This temporary barrier prevents the bees from squabbling and allows them to get used to each others' smell while keeping them apart to start with. In a few days the bees will nibble away the newspaper – you'll know this is happening when you see discarded shreds near the entrance – and the hive will gradually be united. I start them off by making a tiny nick in the corner with my hive tool and then the bees will chomp their way through; you will often find a creative artwork of shredding when you return.

When the weather turns colder and the trees and plants are offering no more seductive nectar, bees love to rob honey. This is dangerous for the beekeeper. A hive can be burgled by bees at any point throughout the season, and as an urban beekeeper you need to keep a constant lookout for these thieves. They arrive in large numbers, hanging around together like youths, with their legs dangling behind them – this signifies that their honey sacs are empty and they are on the prowl.

Weak hives, unable to fully defend their larder, are particularly vulnerable to invasion by bees from neighbouring hives; a huge amount of infighting can result in hundreds dead. As bees will also happily pop into hives whose occupants are dying or have already died because of disease, it is a way of brood infections spreading easily. I close the entrances of weaker hives to the bare minimum, shoving in foams to restrict the entrance gap; if they're on an out apiary, grass will do the same job – it's a bit unsightly but effective.

Searching for sweetness

Stored boxes are also susceptible to scavengers, especially if they are wet with honey. In fact, anything sweet can be a draw for bees at this time of year. Last September, my Fortnum bees flocked around some sticky recycling bins near the store. Staff were very understanding as I hosed down alcopop bottles to remove the sugary remains that were causing the problem. While I don't think the bees were tipsy, they had clearly developed a taste for the hooch as thousands were backing up waiting for their turn.

Fortunately, I have never had a swarm at Fortnum's. If there is ever one, I think it's bound to head the way of the Cavendish and guests will get the rather unusual treat of a beekeeper clambering at height to retrieve his bees. I often get guests hanging out of the windows that overlook the hives, taking pictures and asking questions; it's a rather bizarre experience and I think some love it!

Often in the cooler, more barren months, as bees have to work harder to gather what they think is nectar, I find some rather strange colours within the brood nest. Pink honey has to be the most bizarre. It smelt like bubblegum and I never did find out what it was made from; it may even have been bubblegum if the bees had found some of the sticky stuff to forage on. A few years ago, at the Ludlow food festival in Shropshire, I found a beekeeper selling bright red honey. I couldn't believe he hadn't cheated in some way until he showed me photographs of the frames being extracted. His bees had been foraging on rotting fruit near Hereford. They had ignored the orchard where they were supposed to be pollinating fruit trees, and flocked instead to a nearby strawberry farm where they found fermenting strawberries that had been ruined by bad weather.

In London, I worry about my bees causing a nuisance when they are scouting around for sweet stuff. And you should always be conscious that your bees might be out there causing trouble with something sticky and fruity. In order to prevent foreign bees from invading my space I invest in some of those 1970s-looking brightly coloured plastic ribbons that hang down from the doorway – but not sticky flypaper.

Do your best to keep your extracted wet honey boxes covered or sealed and keep an eye out for businesses that are not keeping their yards free of gloop. Hartley's used to have a jam factory in Bermondsey but the building has now been converted into luxury flats so that's one less thing to worry about. The Tate is very good at keeping all its waste behind giant doors. I've never seen bees on the prowl there.

Despite concerns about pilfering bees, for most beekeepers, September is a restful month. Hive inspections need not take place too regularly as the risk of swarming is small – once every few weeks is fine. If you notice that your hives feel exceptionally light after you've harvested honey, it might be worth building up its occupants by feeding them some sugar syrup.

A new home at last Today, as I cart my first honey boxes into my new studio, deep in thought about how and where I'm going to sell my supply of honey, I notice a honeybee making its way in through the office window, attracted by the sweet smell of spilt and bruised honey.

This bee's abdomen is golden, which suggests it's a lazy New Zealander, prone to scavenging, and certainly not one of my lovelies. It is on the hunt for some of my precious honey. But a bee is a bee so I turn off the insectocutor, which effectively disposes of wax moths and other flying insects that would otherwise end up lodged in my produce, and I escort the Australasian intruder off the premises, using the internationally recognised method of bee removal – a glass and the BT phone bill. (Why is it always the first bill to arrive when you move into somewhere new?)

After weeks of searching for somewhere to live and use as combined honey HQ, I have settled on an old tannery off Tower Bridge Road. It's a vast open space with fantastic light and views to the south, and being only two streets away from my old Bermondsey flat, where Ned lives with his mother, it's an ideal location. It has only one major drawback, in fact – it is on the third floor and there is no lift as such, just six flights of stairs. Instead there is an amazing old-fashioned hoist, with a rather hazardous basket that hangs over the street…. Hauling honey up this way could be perilously entertaining, and I expect I'll end up calling on emergency assistance from friends and volunteers.

Honey in jars or combs For me, the challenge is processing a large amount of honey – most years, I make more than two tonnes – while ensuring that each honey box is handled as little as possible. Weighing

199

25 kilos, each honey box is worth a small fortune, and needs to be treated delicately. With the slightest of knocks, the comb can get detached from the frame, which means that honey will instantly start oozing out of the box. In the past, harvesting helpers have inadvertently damaged combs with innocent but inexperienced handling. These boxes are the result of my sweat and toil over the year and I don't want to find myself wasting it by being careless.

Once in my flat, I need to extract the Fortnum's honey in the next week or so – they're keen to get jars on the shelves quickly. Then I'll give the sticky frames – we call these wets – back to the bees to clean up ready for the spring by gobbling the remaining honey. Returning these to the hive during the day can get the bees rather excited, especially if the outsides of the boxes are also sticky, so this is a task I do at night. Each box is labelled with the site it has been removed from, so it goes back to that apiary, to avoid any cross-contamination. This is the only time my bees are given their own honey back, as it were.

The first job is to separate out the old darker combs that I've used before and which don't look very attractive from the white, pristine-looking ones made from a softer wax. The former will be extracted and put into jars, while the latter will be perfect for my honeycomb. On toast, so long as the wax is soft and mild, honeycomb is delicious. Each comb is individual and I can only guess as to the type of honey beneath the delicate white wax capping; what I do know is that each will have its own special character.

I search online and find an ice-cream factory in White City keen to help. At £6 per pallet per week, the price is cheaper than any other cold storage units I have used; the only decision to be made is to whether I want to chill them at around 0 degrees centigrade or freeze them at minus 25 degrees.

I'm told by the factory owner that they have had some rather strange requests for chilling and freezing things before, but never honey. Last year, they turned down a Channel 4 programme wanting to chill a crocodile they were going to dissect. I am given strict instructions that there must be no bees flying around – I don't bother to tell him that at the temperatures he's quoting bees will have no chance of doing that.

Wax moths

I opt for freezing my combs for two reasons. One is to stabilise the honey so that there's little chance of it crystallising. The other, perhaps more vital with London honey which is, in any case, fairly stable, is that freezing is the most effective way of preventing any wax moth development.

Like the dreaded clothes moth, the problem with an infestation of wax moths is not the moth itself, which won't touch the combs, but the larvae which will munch straight through them.

My Buddhist tendencies go out the window when I spot these grubs, which are quickly despatched between my fingers. At the North London site, a friendly robin appears from nowhere when I arrive there, darting around my feet and waiting for one of these plump delights.

Older beekeepers tend to use mothballs and other chemical products to control moths, but I find the smell takes ages to leave the combs and anyway I like to use as few chemicals as possible. In the past I've spotted both large and small wax moths feasting amid my combs; both wreak destruction, but the larger ones seem to bore into woodwork and damage hives and honey boxes too, so that's the one I dread seeing.

Should you get an infestation, you may find you need to use a blowtorch on the woodwork to get rid of all the larvae. By this stage, your combs are likely to be a lost cause so it won't make matters any worse.

Finding wax moth larvae in combs for cutting is very worrying as they can destroy them very quickly, and educating folk about the wonders of raw comb has been the mainstay of my bee business – but to discover a maggot in a piece of comb already cut would be disastrous. To avoid this I choose to chill and store combs in cold storage facilities at freezing temperatures; this kills off any beasties' eggs.

Going corporate?

As for the last of the honey that I'm going to sell in jars, I've decided this year to extract it even later than usual. Its consistency seems fine so I leave it sitting in honey boxes; besides, I've got other fish to fry.

For one thing, I have a meeting with the food giant Marks & Spencer. Alarmingly, they suggest purchasing all the honey I can produce. In terms of making my honey available to all of London, this could be a step in the right direction; but I'm nervous about jumping into bed with a supermarket, not to mention the possibility that my product will move into the mainstream.

As well as the strict guidelines this retailer expects producers to adhere to, a stumbling block could be price, and yet the meeting is surprisingly encouraging. I'm quick to mention that London honey is an artisan product that in turn demands a premium price. This appears not to be an issue; they still want to secure every single drop I can produce.

I'm keen to see if they could help fund new larger sites for my bees, but the idea still needs more thought. My honey would be sold under the M&S brand and I'm cautious that I will be selling out. Supermarkets have such a lot of control over consumers and are well known for squeezing small producers, so although the offer is appealing for reasons of cash flow, I'm not sure I feel entirely happy with it.

All the little delis I currently supply are admirably local and sustainable; I enjoy working with them. But if I were to sign the deal with M&S, I'd have to let them down. If friends and fellow bee farmers have any criticism of my business, it is always the same: that I don't think big enough. I am in danger of keeping it as a cottage industry and not a brand through which I can launch other products and ranges. But the fact is that I feel more comfortable with beekeeping on a small scale.

If nothing else, my meeting with M&S shows that my bee work has not gone unnoticed by mainstream buyers. In the end, I ask if I can sleep on it and let them know where I stand in a few weeks. But in my heart I think I've already decided to say no.

Just as importantly, M&S is keen to look into bees on the rooftops of its stores, starting with its flagship store at Marble Arch. This could be a chance to fulfil my aspirations of getting bees onto rooftops in every borough as there are dozens of stores across the capital. But again I worry that I'll be massively overstretched by all this, and the fact that their health and safety honchos are already working on plans about what to do if customers are stung instore, though perfectly valid, makes me wonder if this is all going to be too corporate.

I have come up against this sort of concern before and it is important not to be blasé with a quick response. Some clients are very organised and have all the relevant bodies and parties at these initial meetings to discuss all the implications. Others need a more gentle approach and I put on specific talks for staff to discuss what to expect and the amazing benefits of having bees on their buildings.

Upstairs
Downstairs

One of the small independent delis that currently stocks my honey is the Deli Downstairs in the East End. Theo Fraser Steele is a former actor and now scriptwriter, who also manages to find the time to run this neat deli with his wife Sarah. As soon as you enter through their old-fashioned shop door, you are welcomed with some amazing delights. Don't go in there with an empty stomach....

I had been frequenting their store for some time as it was not far from the actress's house and I would pop in for picnic-lunch bits – I'm particularly partial to their sausage rolls and the mackerel pâté made by Theo's mum. Our 'bromance' blossomed when one day Theo passed comment on my work wear – I expect he thought I looked like a tramp – and asked what I did. By the time I left, he'd agreed to stock my Hackney honey.

Weeks later he's at mine, kindly knocking up a honey treat of Baklava French Toast – his own take on the classic snack. This version is quick, delicious and the girls from the studio love it. It's all being filmed in my new studio flat for TV la la land – a BBC2 pilot about bees. Yeah, that's right, we eat like this all the time....

Baklava French Toast

We read about this dish in an article on Australian cafés. There was no picture or instructions, just the name, which was enough to get us interested....

This recipe makes very good use of honey on the comb, which not only tastes fantastic but looks great too. We used Steve's honey from Hackney Marshes.

5 tbsp honey
1 tbsp water
zest of half an orange
a handful of pistachios and walnuts, chopped
a pinch of ground cinnamon
2 slices of good bread
1 egg
1 tbsp milk
2 drops of good vanilla extract
a pinch of salt
butter, for frying

To serve:
a small square of honey on the comb
orange zest
mascarpone

Simmer 4 tablespoons of honey with the water and orange zest in a small pan until it thickens to a syrupy consistency, then take from the heat and remove the zest.

Mix the nuts, cinnamon and the last tablespoon of honey in a bowl. Spread the nut mixture between the two slices of bread.

Mix the egg, milk, vanilla extract and salt in a shallow bowl. Dip the sandwich into the egg mixture. Melt some butter in a pan and fry the French toast until golden brown on both sides.

Serve with the honey syrup, a square of honeycomb and a quenelle of mascarpone.

Packing and labelling In Crouch End, a local Budgens store has already dedicated its flat roof to vegetable production for the community and they are planning to install bees. For now they have ordered my honey to sell, although they are not so keen on my luggage-tag labels and want barcodes on the jars. I have never bought a barcode before and the experience is seriously complicated. First you have to buy them, then you register them before making up your labels.

Fortunately, the ins and outs of supplying barcodes will not be a part of most beekeeepers' autumn. Back when I started out, I gave away my first crop to everyone in the building, which made things easy, but if decide to sell yours you will need to comply with the strict labelling regulations that are in place to prevent consumer deception.

I still know beekeepers who sell their honey with no labels in order to get around the law. I like this idea in theory, but actually it is there to protect both consumers and producers; there have been cases recently of foreign honey being sold as British.

No matter how big or small your production, giving honey away as thank-you presents is particularly rewarding. Friends, family and neighbours will always be delighted – the only possible exception I can think of is dentists.

Honey is not so popular among the tooth profession. It's a known fact that beekeepers have terrible problems with their teeth as they are always sampling honey, especially fresh from the hive, and all those complex sugars play havoc with tooth enamel. My own have acquired a large quantity of fillings due to constantly licking my fingers during the day.

As a commercial beekeeper, I must work out how to package and sell my honey and how to make enough money to take me through the barren winter months and into the spring. It's time to start thinking about Christmas and how a crop can be converted into sales as well as presents.

When it comes to packaging, I've always been frustrated by the lack of imagination behind most honey jars in the UK. I have found it to be stuck in the 1940s – unappealing and drab. In Europe, especially in France, Spain and Italy, honey is a premium product, demanding good prices; there it is given the sort of packaging it deserves, reflecting the effort and devotion that have gone into its production.

Right from the start, when I ran my first stall at Spitalfields Market, I tried to change this by putting my honey in wildly expensive Kilner jars. When you package your first jars of honey, I'd recommend seeking inspiration from artisan producers rather than choosing the labels from the equipment manufacturers. Keep it simple. I've seen some terrible designs: country-life type scenes depicting vicars and tea parties and reflecting this tranquil rural life that us beekeepers are supposed to have.

Just recently I've taken up an offer from a design company in Old Street that bought some honeycomb from me last Christmas. It has offered to do my packaging for free; in return it will develop my range and use this to obtain new clients, which is fine by me. Initial designs for the jars are enlightening and fresh but printing still remains the biggest cost for me to cover.

A few years ago, I designed a hexagonal tasting box with the dear help of some close friends, featuring six different UK honeys, inspired by test-tube samples I saw in New York. This was a big success – I sold several hundred – but they were time-consuming to construct. This year, I'm hoping to return to farmer's markets and increase the number of delis I supply. The larger stores already know their Christmas lines, but at this point in the year it could be good to introduce new products that will be recognised as ideal gifts for the crazy consumer-led months to follow.

Relying solely on an income from honey sales is no longer an option, especially when the season is poor and bees suffer from disease – any savings can quickly disappear. I intend to look at other products I can make from honey and beeswax and try to resurrect old products with a modern spin. The Tate is keen to develop products with me and I would like to try soaps, lip balms and candles.

Flogging your wares

By the end of the month, London's markets will have picked up their pace. Summer is the quiet season with so many people on holiday or out of the city, but between autumn and Christmas it is a golden period to flog your wares.

Often considered the simplest way of selling produce, succeeding at markets requires a huge amount of dedication. You can't just turn up and sell any old shit. You must listen to your customers. You must also make sure the product you are selling has been carefully tested – this sounds obvious, but a beeswax candle of mine nearly burnt down a customer's house once after I put the wick in the wrong way round. The lady came straight back to me for an explanation and fortunately I was able to placate her with some free honey.

I have seen some interesting honey-related products that have not stood the test of time, particularly gimmicky items such as teddy-bear-shaped jars and fancy tins of beeswax furniture polish. In a world where everyone's furniture comes from Ikea, who really needs it? Experienced traders have little time for those not prepared to take selling seriously.

My golden rules for selling at market

- Never sit down.

I think I have once, when the World Cup was on and the market was dead.

- Always look busy.

I take blank jars to label at market, partly because I have run out of time in the week, and also because it keeps me occupied. It doesn't pay to look as if you're slacking.

- Remember to frequently check the front of your stand.

It's very important it looks appealing to customers so your honey is shown off at its best.

- Don't get complacent.

You won't make the same amount of money every week. It will depend on weather and lots of other factors, such as bank holidays and key events that can keep customers away.

- Be nice to the other traders.

Take it in turns to fetch the tea and look after their stalls if they need a break.

- Offer consistency to the consumer.

Turn up to market with a good product week in and week out. Don't change the price to suit a customer – but do give honey away to the local tramp and those with not quite enough in their pocket.

- Expect it to be exhausting.

I find I can hardly get from the sofa after a good market day.

- Be chatty, but don't let this slip into overfriendly cockiness.

- Look hygienic.

Have clean nails and smock or apron. I have walked away from stalls where traders looked like they needed a good wash (which is rich coming from me).

- Finally, have fun but don't ever forget that this is not a hobby.

As soon as you do, you will lose the respect of other traders who rely on this for income.

After Spitalfields, my second market was in Pimlico, part of a network called London Farmer's Markets. To qualify, you had to be producing within 100 miles of London and be inspected by their dedicated team. I became their first honey producer and I still have that stall today at the same location under a giant plane tree in Pimlico – this shields me from the rain in the winter and throws amazing dappled light through my effervescent jars in the warmer months, making them come alive. I love it.

My third stall was among the humongous throng of Borough Market. It took several years to get my pitch and I had heard some terrible tales of politics and infighting. I was turned down for my first four applications because the powerful trust believed honey was not an everyday food item and as such would not be consumed and bought weekly. This is clearly rubbish. My grandmother would eat a jar a week, and there are many other serious honey nuts out there.

Simon Hughes, Liberal Democrat MP for Bermondsey and Old Southwark, even penned a letter of support after the failed applications – no sweeteners exchanged hands. Each laborious application required samples of everything to be sold and a detailed breakdown of why your product was worthy of a place on this ancient trading ground.

Eventually, back in September 2002, I was offered a stall on Saturday mornings, in the shadow of Southwark Cathedral, underneath the railway bridge. I was excited to be offering a product that was made within a stone's throw of the market. I believed that it was going to be the bee's knees and the making of my business.

For the first few months, it was both fun and profitable. I met some great friends and there was good trader camaraderie. My stall neighbour was Richard Hayward, a third-generation oyster fisherman from near Colchester and now my son's godfather. This giant of a man, with huge tombstone teeth, is a gentle soul. He would buy me tea, and when I was busy with bee work and running late he would help set up my tables.

He also taught me how to open a native oyster without jabbing the knife into my thumb, and just as importantly, how to mix a mean shallot vinaigrette dressing for his exquisite delights. There were numerous small producers back then, mainly from farms trying to diversify instead of just selling milk and meat. The day's highlight was trading goods with other producers – cheese, chillies, ostrich meat and cake. Bartering without money exchanging hands is still something I love today at a market and I'm fortunate that people always seem to want honey.

There is a dignified code of honour among traders to take care of others, looking after stalls when you need to grab some tea, and generally helping each other out. I once accidently hooked up my van to a stall selling yoghurt drinks and dragged it through the market. Everyone was waving at me and I was inside the pick-up marvelling at how friendly market vendors are. Fortunately the stall's drinks were in plastic bottles so there were no breakages and the stallholder was very understanding.

Providing it was not too cold, I would take bees in an observation hive to all three markets during those early days, and they always brought crowds to the stall. Markets are great for spreading the word, even if I do get a little weary of being asked the same questions. The most popular being: Do you get stung?

Then my marvellous son was born, and I had to sort out some kind of career plan and buy some pyjamas. Suddenly my Morris Minor pick-up no longer seemed practical – you couldn't fit a baby seat inside.

I had also become disillusioned with Borough Market. It was going the way of Camden Market – very political about how it was run, and full of tourists, only interested in getting a quick bite to eat. For a time the novelty of London honey meant it sold well, although that was not the main reason I had gone there. Back then it was the place for a small producer to sell their goods and meet others, but I felt the market had lost its soul. After eighteen months it was time to leave. I sold the ageing Morris pick-up to a Tube driver and bought a family car.

If you ever produce enough honey from your bees, I can heartily recommend selling your crop at market. There are a few guidelines you need to follow, but sometimes beekeeping associations group their resources and run stalls at festivals and events, and this would be a gentle introduction into this rewarding way of selling.

The bee cab Another brilliant thing I stumbled upon for broadcasting my bees was my bee cab, designed two years ago for Pestival, an arts-based insect festival which takes place at the Royal Festival Hall on the South Bank. When asked if I could contribute anything to the event, I drew a bee-shaped taxicab on a rough bit of paper by the phone. Having had a London black cab – and though it was probably the worst vehicle I ever owned – I knew it was a design perfect for transformation. Plus there are some great similarities between London's cabbies and my bees, both buzzing around the capital, grafting away.

The cab was eventually converted by artists from my sketchy designs, with the support of the Wellcome Trust. It had a mini cinema with a widescreen TV in the back, and an observation hive in the passenger window. Covered in black and gold fur and with some spongy compound eyes on the bonnet, it looked a picture and was a huge hit. Inside, we showed bee films that I'd made about honey hunting in Bangladesh, harvesting honey in Zambia and beekeeping in London. Although it was designed to carry five when on the move, at one exhibition I counted fourteen children in it, plus two on the bonnet tugging at its drooping antennae.

It sadly now sits in a garage in North London having failed its MOT. Despite numerous efforts to try and wrestle it back to life, it remains there gathering dust. I had great plans and wanted it to make visits to schools and businesses, spreading the good word and showing 'B' movies in the back. But it's also stuck in a political custody battle. Because it was supported by the Wellcome Trust, it partly belongs to them and it's questionable whether I have the rights to use it. This has cast a shadow over its future use. If I can find some more funding, I'm keen to build another one. I like the idea of a fleet of them spreading bee propaganda.

An alternative would be to convert my Vespa. It could easily be made to look like a bee (despite vespa being Italian for wasp), but it is currently in the Cotswolds having a sidecar fitted. It was an embarrassing morning when I took it to the company that makes sidecars, and was surrounded by the tough bikers who work there, all staring at my Modtastic classic. Fortunately they loved the ageing beast and the tattooed hairies soon took turns whizzing around the car park on it.

The plan is to have a flatbed on the sidecar buggy onto which I can strap beehives and equipment. I'm hoping it will still be nimble enough to weave in and out of the traffic and avoid the congestion charge. On the drive back to London I stop at a village market where I find a basket maker selling giant laundry baskets, and this triggers another idea. I could mount a basket on the sidecar to transport my hives. I already own a vintage hotel laundry basket with a lid. It would be perfect – just like Wallace and Gromit.

What all beekeepers should be doing:

- As the season draws to a close, it's important to check that your bees are completely ready for any cold snap. Put in hive entrance reducers and mouse guards, and check that weaker colonies have sufficient stores.
- You might want to combine two weak colonies to give them more chance of making it through the winter.
- Consider making hive crown mats from builder's insulating boards. With their silver liner and compacted foam they make fantastic insulators on top of the hive's feeders and will help to keep the bees snug over winter.
- Think about whether you might sell your honey at markets.

OCTOBER

THE LONG MYND

On the first weekend of October, an event takes place in Shropshire that involves fifty miles of sadistic marching up and down Shropshire's bracken- and heather-clad hillocks. The Long Mynd Hike requires a full day and night to complete; it is not for the faint hearted.

By this point in the year, I am ragged from endless bee toil so I have never been tempted to participate, even though I know the hills intimately and friends are always encouraging me to give it a go. Bee moving is not about brute strength – I'm wiry with the unhealthy ability to endure weeks of exhaustion – but I am conscious I'm not getting any younger and my body is already showing signs of years of bee toil, fuelled by cheap and rich cakes. I'm not sorry to have the help of Tom Bean, who always comes along for the jolly at heather harvest time.

Before the race starts, I need to move all my bees off the Shropshire moor. A makeshift combined café and checkpoint spring up – which used to be manned by my father and his chums – to administer sweet tea and scalding soup to exhausted participants. Walkers and runners reach its sanctuary throughout the night, guided by fairy lights and blaring disco music, and my bees will be right in their path at mile 45 of the total 50 of the ordeal.

It would be unfair to expect the participants not only to negotiate a dangerous bog that virtually surrounds the site, but also to have to deal with beehives in total darkness. To an exhausted stumbling trekker, it could be treacherous as they stagger towards the pumping discotheque.

At 1,600 feet, it is also fatal to leave the bees here over winter where the cold, bleak climate would be more than they could cope with. The queens will certainly have stopped laying at this altitude by now and begun their autumn shutdown. They need hauling out to warmer climes, for the season is not fully at a close yet and there is still hope for mild days to come at less elevated sites.

A ceramicist on the other side of the valley, who used to keep his bees at the back of his studio, would lean a pane of glass against the front of each of his hives, to prevent winds from funnelling into them. I always presumed this would be very confusing for the bees, but I'm guessing it worked because they rarely left the hives through winter, and stayed holed up against the mist and dankness.

I have left bees on the Shropshire moors over winter before, but it has been a mistake. With the wild winds and minimal early pollen, they have not fared well.

My beehives are currently alongside Kingy's, clustered onto old wooden pallets arranged in a horseshoe in an old grazing paddock. In the middle of the field is an old black-painted corrugated-tin hut, once used by wealthy shooting parties who frequented the moors in search of its highly priced grouse. It is now dilapidated and mainly a site for underage drinking. The smart composting loo provided for the Duke of Edinburgh hikers has long gone.

It is the only site on the moor for miles around that has trees, which offer protection and security for the bees. They also mean it can be viewed from great distances and hence ramblers throughout the year gravitate towards it as a trig point. Another huge advantage is that the bees are literally sitting in the middle of the heather bushes here and so they have it on tap.

Ever since the 1970s, I have frequented this spot. Climbing onto the wobbly roof of the cottage and pissing from it was an initiation ceremony for my scout group – 1st Eric Lock troop – and it stands in our family's memory as a national monument. To this day, it's the most likely destination for a Boxing Day walk. It is a welcoming landmark that suddenly rises from the mist as you appear from Ashes Hollow – one of South Shropshire's numerous isolated valleys that meander and rise.

To the north is the craggy Stiperstones range, where the Devil himself reputably resides on a stone chair when the fog is down. This is where I used to situate my bees until the heather became so poor that I felt it was pointless dragging them there any more.

The heather beetle has wreaked tremendous damage on its banks and the government agency responsible has refused to address its demise. I have ten-year-old photographs of this heather in full bloom but now it looks dry and barren. The hills are managed very differently from the Long Mynd.

This year, some of my young queens have mated with golden drones despite my adversity to these lazy Australasians – alas, I can't always prevent indiscriminate love affairs – so I now have a few colonies with these genes. Although their big brood nests could be handy for the heather, that's only if they don't starve through their reluctance to work. Though as the hives here are positioned right in the middle of the heather bushes, even these lazy golden bees that favour fast-food forage joints should be able to manage.

Fabulous fungi Before I start loading up my own hives, I am careful not to forget to indulge in my favourite activity at this time of year: a fungal foray among the pine trees and scrubby moorland.

On the dashboard of my truck sits a variety of beloved treasures, including a skin shed from an adder, dried unidentified flowers, various delicate feathers – including the wing feathers from a barn owl – a bright sinister fishing lure with a treble hook found in a tree on a riverbank, and most importantly, my Antonio Carluccio *Complete Mushroom Book*.

This well-thumbed companion is sun faded and extremely well travelled, but it still offers crucial knowledge for tasty free treats across the seasons and helps me avoid fatal gastro disasters. So far, I have been lucky; I have never eaten a badly identified mushroom.

The cep, the prince of mushrooms, is so exceptionally tasty that the hunt to find them has become a notoriously secretive tradition. Forgive me for not fully divulging their location, but I can say that they generally appear amongst the mosses and scrubby heather on the Shropshire moors.

When I spot one, I get very excited. As soon as I've carefully cut my first of the season, I hold it above my head and perform a small jig in homage to the domed beige beauty.

Later, I'll make cep risotto with my camping stove for myself, Kingy and Tom. I fry the mushrooms with a knob of butter and then use their flavoursome juices to cook the rice. It's a perennial favourite and I try to make it each year before leaving the moors.

Tom's trick for weary beekeepers is a version of Flaming Hedgehogs, from mushrooming and foraging guru Roger Philips. Tom uses hedgehog mushrooms, but if you're not confident about identifying them, the dish could be made with good-quality chestnut mushrooms. Sizzle them in some Calvados – they turn into flaming hedgehogs when you light it – then add some cream. We usually all just crowd round the frying pan, dispense with bowls and forks, and opt for torn-off bits of bread to mop up the rich sauce.

In North London, I find chicken of the woods, a massive bracket fungus; it is bright orange and sulphurous but when cooked in garlic, it tastes like steak – astonishingly meaty. On Salisbury Plain, it's all about the cream-coloured giant puff-balls – these massive abnormalities, which can grow to the size of a football, are great cooked in bacon fat and seasoned well. The capital is good for horse mushrooms, and even morel mushrooms that spring up in the spring amongst the bark chippings in urban car parks.

All these varieties are a delicacy for the nomad in me. Wild harvesting and foraging goes hand in hand with my lifestyle and I keep essentials like olive oil and a frying pan underneath the front passenger seat for any enchanting finds throughout the year – all cooked up on my James Bond-style briefcase camping stove.

In London, I also get huge seasonal joy from blackberries, elderberries and wild garlic. I have heard of sea bass being caught in the Thames at Dockhead in Bermondsey, but I prefer the brook trout, poached from tiny streams with a small fly rod called a smuggler. (I use my own hand-tied flies which I keep pinned into the sun visor of the truck to prevent losses – my favourite is one made from a lion's hair which I got from a stuffed lion in Paris.) Cooked in a pan with butter, these speckled lovelies are delicious.

This fantastic cake made by Lara Bernays, Tom's soon-to-be-wife and granddaughter of the Chichester doctor, needs no introduction.

Lara's Heather Honey Harvest Cake

When my Tom heads off to join Steve and David for the mammoth autumn harvest of heather honey, and to bring the bees back south for the winter, it has become tradition for me to send him off with a super-rich, high-energy fruit cake to last the week. A bit of everything goes into it so when they have no time to stop, eat or sleep they can run on cake! This recipe evolves each year.

85g dark molasses sugar
250g last season's heather honey
100ml sloe gin (home-made if you have it)
zest and juice of 2 oranges
350g Lexia raisins
100g sultanas
140g soft dried figs, cut into strips using scissors
150g dried soft prunes, chopped
175g unsalted butter
2 heaped tsp ground ginger
*2 tsp ground cinnamon (or grate your own from a
    cinnamon stick – it tastes amazing)*
2 tbsp high-quality cocoa
150g hazelnuts
75g plain flour
75g wholemeal flour

50g ground almonds
½ tsp baking powder
½ tsp bicarbonate of soda
3 free-range eggs, beaten
1 large bar of delicious dark chocolate

Preheat the oven to 150°C/300°F/gas 2. Line, grease and dust with cocoa a large, deep, round, loose-bottomed cake tin.

Place the sugar, honey, sloe gin, orange zest and juice, fruit, butter, spice and cocoa into a large saucepan. Heat the mixture until melted, simmer for 10 minutes. Remove from the heat and leave to cool.

Meanwhile, crush the hazelnuts by putting them in a zip-lock bag and rolling – I use my nice chunky rolling pin Tom made me. Then toast the nuts for a few minutes under the grill.

Combine the flour, ground almonds, baking powder, bicarbonate of soda and 100g of the toasted hazelnuts. Add this and the beaten eggs to the fruit mixture and mix well. Pour the mixture into the cake tin.

Place the cake tin into the oven and bake for 1¾ to 2 hours, or until the top of the cake is firm and golden. To check it's cooked, insert a sharp knife into the middle of the cake; a few crumbs should remain on the knife when it's removed.

While the cake cools in its tin gently melt the chocolate in a double boiler. I like to give the melted chocolate a good stir which helps give a shiny finish. Spread it over the cake and sprinkle with the remaining crushed hazelnuts.

Best enjoyed when you have a serious job to do!

Leaving the moor This year we slip discreetly from the moor, leaving it as quietly as we arrived. No trace of our expedition remains apart from the light yellow patches of grass where the hives have been positioned. This random patchworking will soon recover.

Kingy and I consider our honey harvest. It is modest but not disappointing.

All the honey, Kingy's as well as mine, is destined for the good folk of London and numerous Christmas markets. As we descend a 25-degree trail, I check the bees but they are calm, cooled by the breeze. Strapped down with bailer twine, they will be secure with Kingy's proper knots as opposed to my own granny knots. Nevertheless we usually check the load outside the Stretton Springs water factory, at the same time filling our depleted water bottles with naturally chilled refreshment from the free tap outside.

I will obtain a fair price for Kingy's share of the crop; he trusts that I will act with honour and pay him his dues. For him, it's not about the money; like me, he loves the journey our bees take us on along with the chat and banter.

He tells me he is sad that I am spending less and less time in the countryside, which is touching, but he knows I am devoted to my bees in the capital and also to spending time with my son Ned who lives there.

I leave Shropshire knowing I've done my best to make sure Kingy's bees are in the finest possible condition. Each hive now has a newly bred queen that will create colossal colonies for him next season, and the colonies' genetics are well mixed, which is healthy. I know he will nurture them with devotion and chase them down on his bike should they swarm from his handcrafted hives.

As we leave, I harbour a pile of seasonal treasures in a felt hat in my lap: the very last of the year's wimberries, a small wild blueberry. Known only by this name in Shropshire, wimberries used to be collected by hoards of pickers on the sunnier slopes, before being sold in London, but because this is so labour intensive they're not worth selling nowadays. Their intense juice provides a sugar rush after another exhausting day.

I drop the bees back at an orchard near my parents, a favourite spot of mine. It hosts an amazing array of ancient apple and pear trees, from which I'm free to help myself to the gnarled marked fruit as part of my rent agreement – which of course involves payment in honey to a great farmer, who always quizzes me on what to plant for bees on his stewardship sections of land.

These bees will remain here for the winter. As my only bees in Shropshire, they will be a safety net should anything happen to the London bees. In that case I will split them in the spring with new queens to make more colonies for the capital.

The hazards of high-rise life

Back in the city, I'm worried about leaving bees at high altitude. I am particularly concerned about those on the roof of the Tate Modern, the highest of my city sites. At twelve storeys up, they are protected by being situated low down in a well, but should they spiral and venture out, they could be hit by terrific crosswinds. There is little shielding them from the harshness of the elements and they might become rather chilled here over winter.

I move them gently on their stands just a couple of feet away from the whitewashed wall that they have already streaked in the spring with poo. This is to take advantage of the lower-rising sun and thereby maximise any warmth available in the winter months.

Another worry is that these hives are next to huge stainless-steel ducts which tower skywards, belching out warm air from the vast boilers that heat the turbine hall. This is likely to confuse the bees, creating a false environment and making them think the weather is milder than it really is.

I consider moving my Tate hives to the museum's secure storage facility for artwork in South London, but ultimately decide the move will be too disruptive. The bees will have to take their chances where they are this winter. I ensure they are snug by wrapping the hives in polystyrene for insulation and weighing them down with bricks.

In New York, most of the bees don't make it through the winter due to the harshness and remoteness of their sites. The bitter temperatures mean they don't stand a chance and new packages of bees are brought into the city every spring.

David Graves, the beekeeping guru in New York, thinks twenty-six storeys is the maximum height bees can be kept at during the season. Anything higher and the bees won't prosper; they'll use any nectar they gather as a source of energy to reach these extreme altitudes. It is my opinion that rooftop bees also need to be strong and full of vigour; only thriving, monstrous colonies will be able to cope with adversity and survive any losses that they might suffer as the result of their remoteness.

I always try and move matured and established colonies, with new or young queens, into the sites with altitude. I feel they respond best to these conditions and offer the greatest chance of producing a crop of honey – weaker colonies or nuclei will not manage so well.

My bees are now coming in with the final honey flow of the year – it's a chance for them to top up their hives' reserves to see them through the winter (or so they think). I'm hoping they will soon move southwards as the weather cools, back into their brood chambers and away from the honey boxes. Bee-free boxes are much the easiest to remove.

Stinky ivy For now, the North London wood is covered in invasive ivy. The giant redwood tree kills off any small saplings or developing trees – this monster was once a sapling itself, bought by Peter's daughter for his birthday – but it's the ivy that really smothers any new growth. At this time of year, the smell of the nectar is thick in the air as soon as I open the truck door. It has quite some odour.

Because I've still left honey boxes on the hives (not a great idea), my bees have been chomping on the highly prized summer honey and are now filling the huge rainbow-shaped gaps in the newly drawn white combs with stinky ivy honey instead.

I once attended a honey festival in Sommariva del Bosco, northern Italy, where I met a honey-tasting guru who was able to describe each flavour in terrifyingly intricate detail. As well as terms such as oaky, overripe, musty and earthy, her most memorable words were 'damp crotch' and 'smelly underpants'. These best describe ivy honey – it really is most unpleasant.

Like oilseed rape honey, it sets like concrete. I have heard of beekeepers removing it from their hives as it can set in the bees' stomachs, which will kill them if the weather suddenly turns cold. I think even the bees eventually realise it's not that special, as I sometimes find it abandoned in brood combs in the spring.

Should you find your bees make a glut of it, it's hard to know what to do with the honey. Some people collect it and sell it, but I don't think it's worth the trouble. Instead, I save it for giving to the bees in the spring when they are short of stores.

Getting ready for winter

A beekeeper's main priority this month, after removing the honey, will be readying bees for the winter season. This involves checking that each hive has a queen, that it is well enough insulated to keep its occupants warm, that it has sufficient stores and finally that mouse guards are installed. And if you have put any chemical strips in the hives (which I don't do) to treat against Varroa, then you should remove them now.

Unfortunately only experience will help you perfect the art of hefting, or gauging by its weight whether a hive contains enough stores. Be careful if the hive still has untaken syrup as you may slop it down the front.

It's generally better to overfeed bees than let a hive become too light; so if in doubt, before you close up a hive for winter, smear some fondant on top of the frames where your bees can get at it easily. Cover it with some plastic or use a freezer bag so the fondant doesn't dry out and remains moist, but make sure there's a tiny bee-sized gap.

I sometimes give my bees a small treat of sugar syrup just now, but this is rare as in the mild conditions of the southeast they generally produce plenty of autumnal stores. In damper and harsher conditions, the bees might need assistance. I carry old 1-litre water bottles full of syrup around for emergencies – and they are usually found amongst any golden-cross bees I inadvertently have, as their colonies are slow to close down and continue expansion right to the last minute, using any stores they have already procured. This is another reason why I am not a fan of this bee in the UK, and I always try and re-queen any with dark Welsh ladies.

Since a mouse can fit through a space the size of a biro – I was once told this by the actress's rotund uncle, who was big in Rentokil – so the purpose of a mouse guard, which can be bought from most bee suppliers, is to narrow the entrance so only a few bees can squeeze through and not a family of mice. It involves wedging a bevelled bit of wood into the entrance or fitting a piece of slotted zinc.

Mice can otherwise build their nests inside a hive as bees will ignore them while they are clustering during cold weather. Although a family of mice can come to a sticky end when the bees wake up in the spring, I have often found them thriving in a hive when I begin checks the following year. Not only will they have stolen the bees' stores, leaving them to starve, but they may well have eaten creative holes throughout the nest.

So now it's goodbye, Bees, as you might not look in on them again until January. As a commercial beekeeper I tend to put a massive stone on top of each hive ready for winter. It's useful to signify that the hive has been put to bed, and it stops winter winds blowing the roof off. On rooftop sites I use a ratchet strap that loops around the body of the hive to make it all one unit.

This year, the weather is still mild at the beginning of the month and I manage to fall behind on my autumn checks of each colony. I'm busy fetching the last of my honey from the hives and if mild enough I also briefly check each colony for its brood and its condition. The frames should be tight together to prevent brace comb pancakes being built in the spring, and a dummy board put in place to fill any huge chilling gaps. This makes the outside frame more snug and is part of a series of measures to ensure the bees have the best possible chance going into winter.

At this time, the queen will begin to lay the bees that will take care of her over the winter period. These bees work less intensely than the summer foragers and therefore last longer, up to six months, meaning they will still be around when the new generation emerges in the early spring. They are crucial to a colony's survival over the colder months; the hive can only be maintained at a constant temperature if there are enough of them.

Rodents in your honey boxes

Rodents can cause severe damage over the winter and you should always ensure that empty honeycomb boxes are carefully stored. It goes without saying that all rodents adore honey itself, and will munch through whole combs if they can get to your stores before you have time to extract it. But they also adore empty honeycomb boxes – the sweet smell does linger – so it pays to wrap them thoroughly before putting them away for the closed season. I use a combination of plastic sheeting and a giant roll of cling film to keep pests out. Rats are the worst; they particularly love chomping right through and cause total destruction; whilst mice seem to just take up residence, building nests amongst the hanging frames.

Beekeeping courses

For me, it is also time to think about giving my beekeeping courses again, one of the most fulfilling things I do over the quiet months. I have met some inspiring students, who travel from all corners of the country and beyond. Some have come from as far afield as Scotland and the South of France; I've also had day-trippers from the Isle of Wight.

All come prepared for four hours of classroom urban bee chat, followed by a honey tasting. I call these courses taster sessions because they are an opportunity for students to see if beekeeping is something that they would like to take further.

It doesn't take long to fill four hours as there is so much to cover. The skill, I feel, is to not overload participants, but instead offer bite-sized morsels of information and tempting enticements to draw them into the hobby.

Innovative designs

I always talk about equipment as choosing it has become a minefield for any beekeeper, especially now that the pastime has become fashionable and fancy new hive designs and shapes are constantly cropping up. Some are based on old-fashioned hives; others just allow the bees to build combs onto bars that hang down; these are more about keeping bees for the sake of keeping bees rather than for honey production, as the honey is difficult to crop.

A beehive can be as simple as an uncomplicated box. As long as it is dry, snug and secure, then the bees will flourish. I have seen bees take up residence in shoeboxes, buckets and even dustbins. They are quite happy to build wild white brace-comb pancakes from the receptacle's roof. It's just that managing bees with wild combs is rather tricky.

The most controversial and radical new design has been the Beehaus, a plastic beehive made by a company called Omlet that also produces brightly coloured chicken coops. Targeted at the trendy end of the market, the hive is not cheap and has created a stir amongst the older generation of traditional beekeepers who think that all beehives should be made out of wood. They don't trust the Beehaus, thinking that it will sweat and that the bees won't have sufficient insulation to remain warm over winter.

Personally, I love it. Targeted towards urban beekeepers, for gardens and rooftops, it has some great advantages over its more traditional counterparts. It's a very long hive so it gives the bees lots of space which can help to avoid swarming; this also makes manipulations such as artificial swarming easier to carry out as the hive can be split into two.

Every aspect of beekeeping has been considered carefully in the Beehaus, and the difficulties the urban and small garden beekeeper might encounter have been taken into account. It even includes a special doorway that can be inserted in summer to prevent marauding wasps from robbing the hive. Best of all, the hive comes on fancy legs to save your back from pain. That may sound feeble, but as previously mentioned, a beekeeper's back puts up with horrible strain from servicing frames. Anything that can be done to ease this gets my vote.

Honey shows October is prime honey-showing season. This starts with competitions at local association level and gears up to the National Honey Show, which used to be held in London and is now in a hotel in Weybridge.

I have mixed feelings about showing as honey is judged increasingly on appearance over taste. If a jar contains an air bubble or is sticky or the lid doesn't match perfectly, then the exhibit is dismissed. Only after considering these external factors does anyone bother to taste the honey, which I believe is the wrong way round. That said, I have exhibited at local level and won several rosettes for my London comb, and the event was good fun.

New office,
new deal

By the end of the month, I'm busy fitting out the office space in my new bee HQ at the Old Tannery – it needs to be a hygienic environment, despite honey being a low-risk food item. Like me, you will need to consider carefully where and how you pack your honey.

In the past, I've run my office from my bee truck, something made possible by smartphones and modern technology. But with business growing all the time, it's becoming quite a challenge and I need a proper premises. I've built a desk in one corner (when I get my weather station set up I can study the computer feeds here), set up extraction in another and a third, soon to be fitted with wipe-clean stainless-steel surfaces, is for bottling and comb cutting. The boxes of combs sit on trolleys now so they can be wheeled around easily, and the location and date of production of each is chalked on the outside, to prevent any mix-ups.

It's good timing as I receive big news from urban retailer Harvey Nichols which is certainly going to demand a proper office HQ. The shop is keen to launch a new range of honey made by my bees. When I meet with Harvey Nichols, I'm told that the range would be rolled out across their six stores in the UK, plus the one in Dublin, and eventually worldwide. My jaw drops – I've inadvertently entered the luxury food-store business. I'm going to need more everything – more volunteers, more planning and certainly more bees.

At my presentation, I notice that the shop sells a manuka honey from the Tregothnan Estate in Cornwall. This variety has created a storm in New Zealand where the PR machine has swung into overdrive with claims that it can only be produced in Australasia. In Cornwall, the honey is made from a limited supply of manuka trees and comes in tiny 4oz pots, costing over £50 each.

Manuka is not the only honey, however, that demands a high price for its exclusivity and limited supply. In the Middle East there is a variety favoured by celebs, made from specialist ingredients such as ginseng which is fed to the bees as syrup – they are unable to go out and forage freely. It's the factory farming of the honey world as the bees don't get to go outside and it strikes me as completely unnatural.

What is even more exhilarating about my plans with Harvey Nichols is that in addition to the range of London honey, they are also keen to look at honeys from the other cities across the UK where it has stores. This is a chance to roll out nationally my blueprint of urban honey to the masses. It's a real thrill in that it feels almost achievable – things really have come on apace.

Although some small delis from outside the capital do buy up my London honey, Harvey Nichols is going to take this to a new level. If I can't manage it all myself, I will have to source producers from these places to work with – thank goodness for the national upsurge in beekeeping.

I already know that Bristol, where one of the stores is situated, has its fair share of urban beekeepers – one morning when I was visiting the headmistress, who lives there, my bee truck was surrounded with interested bees. I was unaware that my truck would be such a magnet – it must have reeked of honey – and it didn't make me very popular. Perhaps I will start my enquiries here. I really like the idea of satellite sites in different cities producing a range of urban honeys.

Although in reality I am already rather stretched even without the Harvey Nichols contract, when I am asked to offer consultation for a rooftop beekeeping project in Amsterdam, I can't help getting excited. Now just where is my passport ...?

What all beekeepers should be doing:

- Your bees should have built up enough stores for the winter now – heft hives to check for weight. Put in a dollop of fondant if they need it.
- If you've put in any Varroa treatments in August, make sure they're all out before you close up the hive.
- Place a good weight on the hive roof to prevent gusts lifting the roof off and chilling the bees. I use bricks or a heavy rock. This also reminds me that a hive has been closed up.
- Clip away any overhanging branches or vegetation, to make sure the bees will take full advantage of the minimal sunlight and remain free of damp. I use secateurs for small work and a bush saw for thicker branches.
- Collect up equipment that will not be used over the winter and store in a secure dry place, safe from rodents.
- The National Honey Show is held at the end of the month – if you have exhibited within your local association and fancy a challenge, then have a go, but remember there are strict guidelines to be followed.

NOVEMBER

WINTER WORRIES

Snow. Already. Can you believe it? The south-east has just a slight covering, but even that spawns an instant dilemma. The temptation is to clear it away from the hive entrances, but I generally leave light stuff as the bees can still breathe through their mesh floors. Heavy snow, however, which may be coming soon, is a different matter. A very different matter. I remove it quickly as bees can often become confused and rush out into a new bright, white but deadly world. Yet even that involves a difficult choice – no one said beekeeping was simple! – as a thick quilt of snow can also keep them snug while temperatures plummet.

In London I'm struggling to ready hives for the winter – even though this damn season is already knocking on the front door. The honey crop is still on the majority of my hives but the bees in the bottom brood boxes aren't bothered as I clumsily remove the honey boxes which sit above the clearing boards. It means I'm now wearing gloves for warmth rather than protection and I still don't use a veil. Yet I'd advise you to remain heavily tooled up until you fully understand your bees' behaviour and characteristics.

On that note, friends have told me about another Esther mini-drama. As she checked her hives, now on a North London allotment, for winter stores, she managed to get another bee up her trousers and she was stung through her pants. Again. Ouch!

Apparently she was witnessed charging across the allotments, trousers flapping, arms waving. The lesson? Bees cling to you when they are cold and have a habit of crawling up trousers. So don't cut corners on protection. Tuck your trousers into your socks and perhaps wear a smock veil for any quick checks. Clearly your bees will benefit from not being poked and viewed after a store check.

As for my winter preparations, well, my mouse guards still aren't in place and I'm behind with everything. Normally I'd be starting to recycle my old dark brood combs by now. This is harder to manage for the amateur beekeeper, but I hate throwing expensive frames away. I use a purpose-built steel box with about six inches of water in the bottom. Then I place a burner underneath which creates huge amounts of steam, making the old combs drop from the frames. A tricky thing to replicate on a small scale. In January I'll have to give them a sterilising bath of baking soda – you begin to see the attraction for smaller producers of throwing away frames. If you do ditch frames, it's important you destroy them carefully as somewhere there will be bees attracted to them. (In summer an alternative option is to use a solar wax extractor that can melt frames several at a time.)

Anyway, there's no chance of my getting any of this done yet. I'm frantic, and I've only just noticed that some hives already appear to have bad Varroa damage. Back in the autumn they were strong colonies, prospering and healthy on the ivy. Now I'm not so sure, although it's hard to fully inspect them with the cold weather.

Indeed, this very same early chill is intensifying the Varroa menace. During the season my huge family of bees was spread across numerous honey boxes and brood chambers. But with the arrival of cooler temperatures they're now compressed into one box, thereby concentrating the mites.

My only hope? I need the colonies to raise enough healthy bees to protect them throughout the closed season. The next inspection isn't until January when they'll be treated with oxalic acid.

This is becoming a huge personal issue. I adore my bees, I'm desperate to protect them and I would hate them to suffer. At the same time, they're also my business – the main part of my income – and I no longer wish to be reliant on treatments which harness huge and powerful chemicals.

You would think that's a good thing; that an organic solution is a positive and enlightening move. But it comes with massive risks. First and foremost it's certain to lead to dramatic casualties. Weak colonies unable to combat the parasites without strong chemical help will fall by the wayside. I could be facing colossal losses.

Some people believe the sacrifice of these more vulnerable colonies is no bad thing. It's naked Darwinism. The stronger, more resilient bees will be the ones left to breed – survivors with the right genes, better equipped to deal with disease. It might make good sense but hell, it's tough love, and it doesn't sit well with my beekeeping ethos. I'm playing God and I hate it.

Several institutions are now researching hardier strains of bees more resilient to mites – and that has to be encouraging. I know that certain of my colonies appear far less affected than others, and better equipped to combat any damage, so I now consciously identify them for future breeding.

There's also a growing array of less brutal potions and lotions on the market. I hear of a soot-based powder, apparently developed after wild colonies living in chimneys were found to be relatively mite free. They're all encouraging developments but none alter the fact that you still need to continually monitor your bees for this bloodthirsty freeloading killer.

Packaging honey At my studio, the fruits of the bees' labours are being processed and packed. The hurriedly assembled office area has become a production hub with Mandana organising the diligent cutting of combs for orders. The wooden honey boxes are gently warmed by patio heaters – not so environmentally sound, I appreciate – to make cutting easier, and the evocative scent of fresh honeycomb wafts down the corridors of the Old Tannery.

Mandana's task appears relentless. Volunteers pack the freshly hewn chunks into boxes with newly designed sleeves. Stacked up on old wallpaper tables, bowing in the middle under the weight, they look utterly amazing. For the moment B&Q's finest are just about holding out, but I have some less temporary equipment on the way to help with our packing, including stainless-steel benches on order from a catering clearance company. My old photographic agent has also offered an enormous table with cast-iron legs like tree trunks – no amount of honey will make that sag.

Boxes filled with raw uncut frames are hauled up to the third floor using the ancient sack winch on the outside of the building, then swung in through the loading bay on Pope

Street. It's a hazardous manoeuvre. If the cradle is brought up too quickly or if it jerks, the now-brittle combs can easily detach themselves from the frames and honey will weep onto the streets. This is one task I insist on undertaking myself.

It's also about aesthetics. If the combs swing and bash against each other as they're raised, they'll look bruised and unimpressive when cut; diligence is required to ensure they look appealing and smart for my growing range of outlets.

I'm still trying to boost turnover; emailing, ringing around, encouraging clients to place orders for honeycomb. Some are hesitant at first as they are unsure it will sell, whilst others are more adventurous and after receiving photographs of the packs they Tweet enthusiastically about their imminent arrival.

Open studio I can't lie. It seems odd to be making money from the bees' very existence. At the moment all my profits go back into their welfare; I'm keen to avoid exploiting them to breaking point just to maximise income as those who favour intensive farming techniques appear to do. I'm always aiming for a delicate balance that prevents the bees from becoming ragged and exhausted. It isn't about the money, it never has been.

So what is it about? I guess I'm striving to run an ethical, environmentally sound business supplying produce to a real range of consumers. I'm delighted this honey is now being sold by established high-street stores – it's a genuine thrill – but I also enjoy selling it to locals in Bermondsey who call in as part of our newly established open studio.

It's pure luck that my new base just happens to sit in the middle of the very latest foodie hotspot of Maltby Street. Small artisan producers, disillusioned (like me) with the now giant Borough Market, have formed their own community. We're only open one Saturday morning a month at the moment and use Ned's hand-painted cardboard bees to guide people to our front door.

Open studio is proving incredibly popular. People like viewing the place where the honey and combs are processed and they enjoy my homage to the humble bee, a mini-museum with its old paraphernalia, books and equipment, not forgetting the giant portrait of my grandmother and her bees.

I'm actually a frustrated honey-shop owner. I've always wanted one and looked at amazing art deco premises in Borough Market twelve years ago. Sadly now demolished as part of the Crossrail development, it would have been a retail hotspot, but even then the rent was huge and I think I'd have struggled to keep it filled with a consistently high-quality product. The more gradual organic growth of my business has suited me perfectly, and now, once a month, the studio is the mini-shop I always wanted and a museum too.

Making deliveries Amid the year's last frantic tasks I'm also trying to deliver an extraordinary amount of honey. The order book is swelling from delis across town as word spreads about London honey-combs. I'm keen to sell as much as possible for Christmas and that means more markets, more travelling for deliveries and more promotion.

Delivering can be the most rewarding and satisfying experience, yet it's also incredibly time-consuming getting fresh produce to clients. With such a small team it also takes real planning to deliver it just before the weekend when most of it's sold.

It's a logistical headache. I often pack more on board than I need as orders are now coming in when I'm on the road. I plan a strategic route around the different sites and gladly accept coffee and treats when I hand over orders in scruffy reused boxes tied with string. I consciously try to minimise my packaging and recycle as much as possible, but I also know that honey boxes need to be clean, presentable and not at all sticky.

We all have to start somewhere and I'm keen to be as accommodating as possible if clients ask for credit. I try to find the right balance but I now have much larger overheads and it really focuses the mind. Indeed, it comes as something of a relief to receive an order for 1,000 combs from established client Fortnum & Mason. It takes my army of helpers all night to cut and another day to package up, and it still needs to be carried down six flights of stairs by hand – but it's all one delivery and I know I'll get paid.

Even though I've only just moved in here, I realise that if business continues to increase I'll have to find a new venue at ground level. There's nothing else for it. For now my calf muscles are bulging. And it's not just me working up a sweat. The studio's lack of internal lift means that, much to the annoyance of parcel delivery drivers, jars, cartons and equipment all have to be manhandled up to the top floor.

Senseless vandalism The early freeze isn't November's only chilling news. I've just returned from another round of deliveries to a dozen or so delis when I receive reports of vandalism on some of my hives. It isn't the first time.

Thirteen years ago, just as I was starting to expand my bee empire, one of the first commercial sites was in an old electricity substation. The company which was then the London Electricity Board had replied to my pleas for unused green space to site my hives. I'd received numerous letters from utility companies explaining that their space was in constant use but the LEB lettings department was most obliging. Its representative drove around south-east London with me visiting numerous sites until we finally settled on one near Plumstead.

It was a wildlife paradise spread across several acres and the bees went wild. It was a huge success, producing a toffee butterscotch summer honey along with a light floral-scented flavour in the spring, when a nearby giant laurel bush would shake with bee activity. There was even talk about them providing a free shed and workspace.

By year two I'd expanded to over twenty colonies in old battered hives positioned in long uncut grass. It was a site that appeared to flourish in an unmanicured state. It was a haven for bees and despite the fact that I was relatively inexperienced and flying by the seat of my pants, the colonies were startlingly healthy and virile.

It seemed too good to be true. It was. I naively believed that the bees' protection lay in the site's razor-wire perimeter fencing and the thousands of volts flowing around the site from the functioning electrical facility. Critics had predicted that the bees would struggle with the magnetic pull of the energy, but their demise lay in their ground-level position. They were too easily reached by idiotic vandals with lighters.

It's alarming to find the aftermath when a hive has been torched and burnt to the ground: a few blackened nails and some mangled wire, that's about it. The bees never stood a chance. Once each hive was lit it would have flared intensely like a candle, with wax and honey oozing out through the entrance. Only two survived. They were partially charred and, when I found them one hot sticky June morning, the bees still flying from these heavily listing ships were understandably defensive and aggressive.

I dropped to my knees when I saw this atrocity, and I'm not embarrassed to say that I broke down in tears. I couldn't understand who would commit such a crime or how on earth they'd gained access to such a heavily protected site. The remaining two hives were moved out that evening to my new site in Barnes. Months later one of the surviving hives was diagnosed with European Foul Brood and had to be destroyed by the ministry – the frames were burnt and buried. I've never had this disease before or since and believe it was provoked by the trauma of the vandalism. Although I still have the keys, I've never returned to the substation site.

Fortunately this has been my most colossal loss to date apart from the seasonal losses, counted each spring after winter fatalities. I used to lose a handful of bees over this period but it could be far worse. I heard of some people losing over 60 per cent of their bees over winter while my worst tally is probably 30 per cent. It is, however, something you need to prepare for by making an action plan over the previous season. You need to make new colonies to replace losses and ensure numbers are workable and constant with young virile queens.

Animal vandals

So what of my latest attack of vandalism this November? Not more bored kids with lighters? Well, no, it turns out it's arriving from a more unexpected source. The heavens. Or at least

245

the skies. I should stress I'm a bird lover. I never want to harm a single bird and this year was thrilled to see a red kite swoop past me on the rooftop of Fortnum's. I've also watched terns nesting in East London and a kestrel glide past me near Tower Bridge. It has been a longstanding love affair. As a kid I was a young ornithologist, logging and ringing birds at night as they came in to roost.

When it comes to the green woodpecker, however, I would gladly make an exception and wring its bloody neck. In the summer you sometimes see this sinister feathered tormentor at ground level nibbling on ants. I can live with that. It's just when the weather cools and those tasty treats are no longer available they turn to fast-food joints, aka beehives. They actually vandalise them to gain access to the bees. It's never the other varieties of woodpecker, just the cunning and ingenious green one.

In Bermondsey, I'd often witness blue tits pinching the occasional bee that shoots out in cooler months for a poo or a quick drink of water. But these particular woodpeckers are in a different league. They avoid the front entrance and bore straight in through the side of the hive. The bees, disturbed from their slumber, gravitate towards the hole to combat the intruder, where the cunning invader hoovers them up by the bucketload.

Occasionally wooden hives survive this onslaught and you will see a peppering of holes on the walls – like sprayed machine-gun fire – that are the legacy of botched raids. The nucleus hives, however, don't stand a chance. They're opened up like tin cans with polystyrene strewn everywhere across the apiary. Colonies will struggle to survive this trauma, especially as the little feathered bastard is more than likely to return for a further feeding frenzy.

If you suspect there's a green woodpecker around, then cover your hives in chicken wire. I've also tried fruit mesh and even bright orange mesh from roadworks but in time, the smart predators are usually able to work out how to combat this extra protection, sometimes crawling underneath, sometimes pecking through. The metal wire needs to be wrapped around at some distance from the woodwork so the beaks can't reach in and drill through.

In Malaysia, I've seen giant nets strung up on poles around the edges of an apiary in the middle of the rainforest, deployed to prevent attacks by bee-eaters. This most beautiful of birds has vibrant electric plumage and an insidious appetite for critters. Rather than drilling into the actual hives it scoops up bees in its forceps-shaped beak. I heard last year that a nesting pair had arrived in the north of England. Local beekeepers would have really struggled and wondered why their numbers were down. My only hope is that they are not encouraged to spread south. You'd think the northern English climate would be a deterrent.

Badgers, I've heard, can also vandalise hives. But unlike their African relatives, I've yet to hear of them causing destruction in search of honey in the UK. In Shropshire they clamber over 2-metre-high barbed-wire fences that protect the bees from cattle – cows love to rub their itches against the boxes – but it's only because it's en route to a field of maize, their preferred diet. In North London, near my hives, there's a relentlessly expanding badger sett alongside the fox dens. I know they're there. I've seen the huge ruts in the rich earth where they dig for bumblebee's nests and in the banks where they're after wasp's nests. Yet the beehives remain untouched. Watch this space.

Hives on a rooftop should be safe from most marauders except, perhaps, rodents so you'll need to fit mouse guards.

Any queen wasps bedding down for the winter are easily evicted and squashed.

Snug at eight storeys

Up on Fortnum & Mason's rooftop I use carefully shaped pieces of oak, which narrow the entrances down to a tiny bee width to keep the mice out. So far they've never been hassled by woodpeckers.

I also bring their ornate facades indoors, to prevent the wood from swelling and the paintwork cracking in the damp. The double-walled hives offer great insulation over winter despite their eight-storey altitude, but I fill any internal gaps with polystyrene boards to ensure the bees are even more snug inside. These hives use large hooked rods that connect the roof to the floor inside, and there is little chance of them blowing over as they must each weigh several hundred kilos.

A dollop of sweet sticky fondant is placed on top of the frames as a final treat for these high-profile bees. As yet they've always fared well in their luxury penthouses, but I will come for one last check next month.

A restaurant delivery

My nephew Luke is with me now for work experience. He is a jovial guy and at fifteen a bit of a prankster but he's good company as I undertake these final checks. I send him on a mission to deliver honey to Mickael Weiss, a fabulous chef. His restaurant is Coq d'Argent in the financial City. I met him on a cookery programme last year and he has been using our honey ever since. He orders three to four frames a week from which he makes up a variety of dishes. He emailed me this recipe for a simple but amazing dressing.

Spiced London Honey Dressing

Hi Steve,

I've always loved fresh honey and was fortunate enough to have a couple of beekeepers close to where I lived as a young boy. Fresh honey on French toasted baguette and butter was one of my favourite after-class snacks.

The London honey has been on my menu ever since I discovered it a couple of years ago. What I love best about it is the complexity of the different flavours and texture you can get from honey depending on its location.

It can be used in its comb form, just a small piece on the plate with cheese, in a dressing, as a last-minute glaze or to just finish a sauce. We also use it in dessert.

Simply no better honey on the London scene at the moment than Steve's.

½ tsp turmeric
½ tsp cumin
½ tsp ground ginger
½ tsp star anise powder (or a whole star anise)
½ tsp curry powder
1 lime
1 lemon
50g sugar
150g honey
100g soft butter

Roast the spices on a flat tray for three minutes at 180°C/ 350°F/gas 4.

Zest the lime and lemon, extract the juices and place in a pan with the sugar. Boil until the mixture has reduced to a thick syrup with a golden colour, then pass through a fine sieve. Add the honey and the roasted spices and bring to a light simmer for two minutes. Leave to cool to room temperature and incorporate the butter.

This is fantastic served with roast pork, pigeon or chicken. You know this works, Steve, you have tried it enough times.

Mickael

Of course my nephew manages to get lost on this first task of the day, delivering to Mickael's restaurant. It's his first time on his own in the capital and what should take him twenty minutes from the studio ends up taking three hours – but hey, he's seen the bright lights and met a fantastic chef.

Cosy clusters Now, as the temperature drops around hives across the country, bees will begin to cluster in a tight ball around the queen, keeping her snug in the centre of this warm mass. If she's still laying, the rate is drastically reduced and bees are employed as heaters to protect the small amount of brood from the cold.

They burrow headfirst into cells and vibrate their bodies to radiate heat through the combs, while other bees will shiver and shake to maintain a constant heat throughout the cluster. At the same time, there's minimal movement and wasted energy so hopefully, with sufficient stores built up, their consumption will also be reduced over the coming months.

They've now become a fragile colony which needs to be left alone. Its very existence is beyond the control of anyone else. You have done the best you can for their health and safety, and hopefully you'll see them flourish again in the spring.

Sustainable food

This month I'm invited to a meeting with Sustain, the alliance for better food and farming. It's a great organisation that eleven years ago was supportive of my crazed ideas about urban beekeeping. From its early hippy vibe, it has expanded to manage a variety of incentives for local and national government as well as its own projects. The Capital Growth scheme has been hugely successful with hundreds of small bodies, schools and communities applying for grants to start their own food-growing groups, spawning vegetable plots across the city.

Now they have a potential new ally: the Mayor of London. This meeting is my chance to learn more about his plans to fund the placement of fifty beehives across the capital as part of a further community and educational project.

Some beekeepers regard it as an alarming development, fearing a new range of beekeepers hungry for instant knowledge, in a city which is perhaps already exhausted; but if managed well I believe it could be highly successful. My biggest reservation is that it could be political with the mayor exploiting the increasingly popular image of bees and the recent growth of beekeeping as a mainstream trend.

I'm delighted to report that the meeting is very promising and I make it clear I'm keen to become associated with the project. It's important to ensure that suitable trainers are employed to teach these new London communities and that the groups fully appreciate the tremendous responsibilities that come with keeping bees in this massive capital city.

Fortunately there is no longer any shortage of people who can teach to a high degree of competency, although some new entrants to the craft clearly feel they're instantly extensively knowledgeable and able to offer training at extortionate cost. I hear that one has even crossed to the dark side and become a bee inspector. If they only have limited bee experience this is a worry – it could certainly make my life harder.

To be fair, inspectors are worthy adversaries, if sadly under-funded and often unpopular. Beekeepers are encouraged to register their hive numbers and locations with the ministry but it's not a legal requirement as it is in New Zealand for example. It should be. In this highly populated country and indeed city, with so many hives in such close proximity to each other, disease is on the increase. It's a far cry from the days of my old bike trips around Bermondsey when my bees' rarity made them easily identifiable and it was very easy to see where they were foraging.

This season has been different. Very different. I've noticed a colossal rise in interest and activity – ranging from regular answer-phone messages from inexperienced beekeepers wanting advice, to the dramatic increase in the number of swarms I've witnessed around town. I know at least one badly managed site which produced several swarms this year. Unfortunately the site was in the middle of Soho and the swarms were seen by hundreds of commuters, many of whom filmed it on their mobile phones.

The partnership with the mayor is being launched next month at the Royal Festival Hall as one part of a massive seminar on London bees that includes a tasting bar for local honey. I expect a great attendance and a good mix of mature beekeepers curious to know what all the fuss is about – after all, they've been keeping bees for years – along with recent recruits. It should be wonderful fun. I can't wait to see the new faces of urban beekeeping.

What all beekeepers should be doing:

- Start reading up for next season.
- Make occasional site checks to see that all is well with your hives from the outside.

DECEMBER

TIME TO RECUPERATE

When I was a commercial bee farmer working in the countryside, I used to see the next few weeks as a chance to relax, recoup and gorge myself stupid. Holding back on my annual holiday until now wasn't just normal, it was essential. The bees demanded my every waking hour. So December heralded the start of a restful time when I could melt back into my lime-green, mouse-ridden sofa and relish a break from the work.

Over the rest of the year I constantly worried about my bees. I'd close my eyes at night exhausted from a day in the field and all I could see were frantic bees scurrying around on their combs. I had nightmares and near-panic attacks. Had I made the right manipulations? Had I fed them enough stores? Was disease about to strike them down?

But enough was enough. Time to stop fretting. Christmas was warm, fuzzy, payback time. Venturing out on crisp, sharp days I'd deliver honey to apiary landlords, partly as rent and partly in apology for having inadvertently driven across crops I'd mistaken for grazing pasture or for my swarms having taken up residence in thatched barns.

And the honey gifts were also a thank-you. A huge thank-you. An acknowledgement for their having planted wild flowers, borage and other bee-friendly plants that my bees relished when pollen was low. And a grateful recognition for their sterling work during my regular driving disasters, sometimes using 4WDs or tractors to salvage my truck from ditches in the middle of the night. Honey was the least they deserved. It was from the land – and it was from my heart.

This was a time to sip tumblers of whisky and sample home-cooked mince pies in local parlours and kitchens whilst discussing the prices of wheat, milk and honey. In Joe and Ann's kitchen it was a chance to roll the dice and be a guinea pig for Joe's alcoholic experiments – sloe gin was my favourite but I was never sure about damson vodka or plum whisky. These were joyous, relaxing times as everyone who rented space in their barns would gather in the kitchen to talk about everything from global politics to Joe's pessimistic take on the plight of the bee.

Their generosity has always been humbling. Ann once presented me with an immaculate Harris Tweed jacket that had belonged to her father – a dry-cleaned Christmas treat. I still have it today in my winter wardrobe and can be seen wearing it on YouTube in a film made for *Guardian* online. I'm beekeeping in the winter on the roof of Fortnum's and have my veil wedged under the collar. It looks sharp but completely impractical – it's somehow perfect for a Fortnum's gent!

And this year, David has offered to give me forty new colonies for the spring. I asked, 'For free?' and he replied, 'Yep.' He explained that when he was expanding his bee business, an old master had given him forty new colonies and it was now time for him to pass on this gift. It is an enormously generous thing to do and I hope one day to be able to continue this tradition with a new bee farmer. I only hope that the profession will still be prospering when my turn comes.

Occasional site checks

By December the bees in these rural areas were beyond my control – they were running the show – and any interference with the struggling colonies could risk their survival. I might visit the more exposed, windy sites to check that the hives' roofs hadn't blown off or that they hadn't been knocked over, but there certainly wouldn't be any intimate bee handling taking place. The bees on any apiary, like me, were now in a catatonic state of slumber – zombies who did not want to be disturbed.

If you have bees on a remote site then I'd advise you to visit them at least once during the month to check they're still cosy and safe. I've discovered hives knocked from their stands – either by stray livestock or by tractors who failed to spot them on overgrown land – with the bees cruelly exposed to the elements. I'd also repair any fences or gates, while ensuring the bees weren't damp from dripping trees and vegetation.

It might be the middle of winter but if you arrive in time a knocked-over hive can sometimes be saved and rebuilt, especially if it has landed on its roof – although any occupants who've survived this ordeal are understandably very defensive in the spring.

They can be pretty resilient to what's going on outside. On a hill overlooking Shrewsbury, my bees would provide a station for posh pheasant shooters as birds were flushed over their heads. Spent shells were scattered around the hives yet the bees seemed untroubled by the noise. They simply ignored the carnage and remained in a tightly huddled ball.

I'll still heft a few hives to gauge the rough weight of stores. Devices are now available – they're still pretty pricey, however – allowing beekeepers to remotely monitor the weight of a hive. They sit under its floor, feeding information via a mobile phone back to your computer. Special software then alerts you if you need to race off to an isolated apiary with honey boxes (if the weight has increased) or syrup (if it's drastically reduced and the bees potentially starving).

It even provides a warning if the hive is opened by an intruder and GPS tracking if it has been pinched – something which I read is on the increase and which is hard to defend against. I make my hives easily identifiable. In fact they stand out like a nun at a rave. They're bright yellow and covered in London Honey Company stencilling. They were already pretty unique as I bought them in Italy, so I'm banking on them being easily traceable should the worst happen.

What other last-minute checks did I make before my December hibernation? I examined my entrances to ensure they hadn't become inadvertently blocked with dead bees, using a small twig to remove any lodged in the gap. A blocked entrance can stop the bees exiting in milder weather and lead to a complete colony collapse. But do take care with that stick. Jabbing can rudely awaken the bees and stings seem to hurt that much more in the closed season when you're less immune to the sharp jab.

Relaxation and reflection Finally, all tasks accomplished, I could now breathe and attempt to get some normality back in my life. It was the chance for my body to recover from the physical stresses and strains I'd subjected it to over the year. My niggles, aches and curved back appreciated the break from dawn risings.

Slumber became a full-time profession in my four-poster bed made from boat paddles with old canvas sails draped over the sides. Its oak panelling was perfect for my Elizabethan-themed bolthole where I'd retreat from the winter chill. The BBC's iPlayer would become my laptop home page, replacing the Met Office's hourly weather forecast.

This was also my opportunity to reflect on the beekeeping season and review what I might have done better. Failed techniques or manipulations were examined, along, more happily, with successful trips to migratory honey sites. I revelled in triumphant yields, crediting myself with my fine strategies but also acknowledging the roles played by good weather and healthy young bee stocks.

David and I would also sit in front of the fire in his tiny Welsh cottage and start making plans for the following year. Huge plans. These military-style meetings would take hours and involve home-made cider and David's signature dish of liver and onions. One of his top five meals cooked in a single pan, it not only saves on washing up but at the right time of year can easily be made in the field.

While on the subject of scoff, I'd favour massive meaty stews containing gamey roadkill deer or pheasants cooked with barley and pulses, followed by heavily buttered rich cakes for dessert. I'd also use the break to fill the freezer with home-made bunny pies, sourcing the ingredients during ferreting trips with Barry the Ferreter on the slopes of the Leaton Estate, just outside Shrewsbury.

Dinner over, I'd crash in front of a huge crackling fire, relishing the opportunity to re-read old bee books and articles torn from newspapers that I'd failed to fully digest and had dumped in the log basket for this very occasion. With time on my hands, I even developed a way of watching TV from the bath. It was a bit Heath Robinson, reflecting the screen off a giant Victorian mirror perched on the bathroom's mixer tap. It was heaven. Even the cats adored it, perching themselves on the side of the bath and swatting my toes as I waggled them to wipe the misting mirror. I'd emerge after several hours of soaking, light-headed from strong honey homebrew and heat, my skin poached pink and wrinkled like a prune.

My hands would lose all the tough skin and calluses, become soft and easily damaged. But most rewarding of all was my trousers. They'd once again fit tightly around my waist. Reverting to the winter notch on my belt was a comforting signal that I was now deep into this glorious period of gluttony and self-indulgence.

Pre-Christmas mania

But that was then. This is now. The idyllic month of recovery appears a distant memory as I prepare for the onslaught of Christmas. The booming seasonal business is my chance to fund future bee projects through the barren months and to generate enough funds for new equipment and further expansion. There is no time for loafing in London.

The studio is now a super-hygienic bottling line, churning out hundreds of golden jars of loveliness from the stainless-steel storage tanks known as ripeners, which can hold anything from 25kg upwards. They often contain a fine filter in the roof, allowing filtered honey to settle with its yeasty scum rising to the top before bottling. They're costly but a very good investment.

I'm not alone in sharing living quarters with bee paraphernalia. Even celebrity types do it. BBC newsreader Bill Turnbull, who visited Fortnum's bees in the summer, told me he stores his honey supers in the living room and spinner in the kitchen. His wife covers it in fairy lights at Christmas as it's now part of the furniture. I giggled at his tale, reminded of storing honey and equipment in a tiny Bermondsey flat – it's something you will need to consider, as you will have realised by now there are key bits of bulky equipment that will need homing.

At the other end of my studio is the assembly tagging and labelling zone. The two are separated by a partition to avoid contamination and I've become obsessed with soap dispensers and hand-washing areas, as well as meticulously recording a log of stock batches. I'm also developing an unhealthy interest in the price guns used to print the legally required Best Before dates and Batch Numbers. Aggressive price-gun fights frequently break out with Bethan my new honey helper, who is considering a career in either MI6 or teaching history in the Maldives – I suggest combining them.

Once again you should be on the lookout for some beekeeping bargains; beekeepers can secure fantastic cheaper 'second quality' items around this time of year. Usually made with more knotted wood than their premier products, they are still fine in quality and more affordable. I use hive makers in Europe who offer even greater value on larger orders, which should arrive at the flat in the New Year ready for my incompetent assembly and a frenzied nail-gun assault.

Christmas markets It's a fabulous month to be in London. Numerous festive markets spring up and it becomes rather tactical knowing which are worth attending and what to stock. Some charge a set price per day, others a percentage of your takings. Unsurprisingly, honey is a huge hit for presents – especially for grandparents and honey nuts. At previous December markets,

I've had to return home before 10 a.m. for more stock, having seriously underestimated honey's popularity as a gift item. Selling out of stock is rewarding but not so early in the day.

Markets are now at their busiest and largest. Stallholders politely jostle and lobby for premier positions and it's a welcome return for the makers of home-made Christmas puddings and cakes, alongside game sellers taking orders for goose, partridge and pheasant. These are happy, evocative events full of atmosphere and charm. I love them. There's more than a hint of Dickensian London about the steaming breath of shoppers on a chilly Yuletide morning.

We sell hot honey and lemon to comfort visitors who cup their hands around the warming brew. But the weather is particularly important for bringing out shoppers at this time. The last Saturday before Christmas is traditionally the highest turnover – and you pray it leaves a good bulge of notes in your pocket.

And wouldn't you know it. This year's weather is awful. Heavy snow restricts the numbers attending a new market at South Kensington and by midday the awning is creaking under a fresh fall. My honey labels are smudged by melting flakes. I'm worried about getting home but still have orders to deliver, so I pack up early and struggle to get my van from the outdoor car park that lies on a hill.

Rear-wheel taxis are now sliding and stopping in the slush up Kensington Church Street. I'm luckier. My front-wheel drive van easily reaches the top, assisted by several hundred kilos of honey and candles in the back. Shop owners are pleased to see me. My work's far from over, however. It's time to head back to Bermondsey to heat up a giant tank of beeswax and make more candles. It sounds romantic, festive and fun, but tonight I'll be doing it alone.

We always sell piles of aromatic candles, made from our own beeswax, on our stalls. I choose simple old-fashioned shapes that are then dipped to build up a unique coned profile. I offer just two sizes, as it's near impossible to fit every width of candleholder – especially fine silver ones in Pimlico.

I'd start by rolling your own simple variety from flat sheets of beeswax – the candles I first sold on my market stall – before moving onto silicone moulds. There are some disturbingly phallic designs – for the hobbyist perhaps – and a terrible mould that produces a naked couple in an embrace. Why anyone would want that on their table I'll never know.

I love the idea of my urban beeswax illuminating the capital's households with a gentle glow. I'm now researching more products that use London beeswax and honey to cleanse and heal the city's workers – after all, Mr Fortnum and Mr Mason started their store 300 years ago trading beeswax in Piccadilly.

Indeed, this year has taught me to be less reliant on honey production for income – it's so damn inconsistent and reliant on a multitude of factors outside my control. I'm hoping the new avenues – including cosmetics – will give me a wider, more secure economic base and transform the London Honey Company into a brand. I know, I know, it won't ever be Coca-Cola or Nike but that's not the point. In the past I've suffered from not thinking big enough.

Old-fashioned attention to detail

So it's hello to large glass Italian urns with carefully cut chunks of comb. They're a fiddle to assemble but look seriously impressive with a gift tag and fancy ribbon and prove a big hit when trialled at the start of December on market stalls. You can often buy them in the UK containing European acacia honey, which is slow to crystallise and remains clear, displaying the comb to its full potential.

David already produces superb giant vessels for Fortnum's, filling them with golden Welsh honey – and they're a huge Christmas success story. I, of course, use London honey for a classy stocking filler.

All this extra work and product development requires additional sets of hands, manically tying fancy bows on jars and gift boxes. Spirits are high, but even the most nimble fingers tire over the final days in the grotto. But what can you do? These fine stylish finishing touches can't be replicated by machines, although apparently there is affordable technology to fill jars and make labels for the hobbyist.

As the month wears on, the Old Tannery is alive with red ribbons and gold tags. Delis that were cautious about buying large stocks of product lose their inhibitions, returning rapidly with reorders after selling out. It has become a struggle to keep up with demand. A welcome one, but still a struggle. I'm forced to draw a line on supplying any new stockists for fear of letting my regulars down.

I pimp several mechanics' flatbed trolleys to move heavy honey boxes – stacked eight at a time – around the studio. To avoid mix-ups, once again each is marked with chalk, signifying the different apiary sites from which they've originated. Builder's plastic cement-mixing trays are screwed onto the base of each trolley to catch honey from weeping frames – a classic beekeeper's adaptation of a pricey option at a fraction of the price.

They're made by Nash, a former boy-band guitarist and now artist. He's a creative whizz, and has knocked up a striking display stand from an old rack of brass coat hooks and recycled pallet wood. The candles hang from the hooks by looped wicks, and a giant bee, made by Henry the black-smith, is attached to the top as a final flamboyant touch. Despite requiring two people to lift it, the stand – highly stable in windy weather – is a dream display for outdoor markets. It all helps tier my stall, something I've considered on slow days as I tend a near-deserted stand. Displaying items at eye level helps sales and I now use old hives to raise jars and baskets for presenting combs.

We have markets booked in every weekend now and orders are flying out daily. I constantly remind myself that no matter how busy we are, each item in every order must remain a quality product. I try to be diligent with each departing box of goods. A huge brown-paper roll is unwound to wrap orders, offering a degree of protection before they're stamped with a bee-print address label and the handwritten supplier's details.

Yes, it's time-consuming and old-fashioned. But I like to think that when someone opens one of my orders and finds their jars neatly arranged with labels and tags facing regimentally in the same direction, it suggests more than the early onset of obsessive-compulsive disorder. It should also reflect the fact that their order has been compiled with care and devotion.

Along with our use of aprons and traditional scales, it's all part of my retro 'brown paper and string' philosophy. I hope it's also reflected in products displayed with manila luggage tags and smartly stamped labels. They're spot on for stores demanding crisper cleaner presentation.

London Bee Summit Away from the sales front line, the long-anticipated London Bee Summit brings some masters of our craft out of the wood-work. It's a chance to catch up with beekeepers I haven't seen for years. After a press launch of the new mayoral bee initiative – Boris's Bees! – on the rain-lashed roof of Tate Modern, it moves on to the Festival Hall. Both events are hosted by the delightful Rosie Boycott, whose father, I'm delighted to discover, used to keep bees in Shropshire when she was a child.

London beekeepers are encouraged to bring honey for an unofficial tasting competition. It's brilliant. There's no other word for it. Just brilliant. The jars are a wild mix of shapes and sizes – a complete contrast to the National Honey Show's strict guidelines – and attendees are asked to vote on the best flavour. I opt for dark honey from Holloway as its taste is unusually deep and unique; the beekeeper is unsure of its source. It wins.

This gets me thinking. As I've so far failed to fulfil my dream of putting bees in each of the capital's boroughs, I'd love to produce a honey-sampling pack of every variety of London honey. It would be a real treat for the obsessive honey lover. I start frantically taking down phone numbers of the beekeepers whose honey I love.

Perhaps slightly high on the samples I've tasted, I deliver a short talk about bees in the capital, explaining how well they've done this year and what great yields they have produced. It's all plain sailing, although if I was to be brutally honest, I'm still apprehensive about their health.

Knocking-off time More optimistically, I've set a deadline. Twelve noon. Christmas Eve. Knocking-off time. The final work order must be sorted as both hands point to the top of the clock. Then that's it until the New Year. I'll head to Shropshire to celebrate, cook dinner with my beloved sister and then collapse in a heap knowing I've given it my best shot.

But I'm not there yet. Not quite. I'm still struggling to find enough experienced people to man my market stalls – and that's critical to a good day's takings. However, as so often in the past, the ever-reliable Mandana steps up to the plate.

In reality, the products pretty much sell themselves nowadays and I'm now holding back on the honey sampling and promotions I might consider on slower days. Instead I'm stocking higher value items including whole frames of honey – carefully wrapped in cling film to prevent sticky leaks. Tied with rich red velvet ribbon, they're quickly proving very popular pressies. On one day, an architect buys seven for his clients.

For all the high-octane excitement, London lacks the countryside's celebration of beekeeping's year end. I really miss it. In Shropshire it would have involved a wassail in the grounds of the big old country estate where my charming Butler's Cottage was located. My paganistic ritual was a rather creative affair, but I think the landowner, farmers and gamekeepers loved the chance to see what the crazed bee man had created for their winter entertainment. We also would serve up hot toddies to Kingy's recipe.

Kingy's Hot Toddy

Bushmill's whisky – a healthy dash
heather honey – a teaspoon or so
1 lemon
cloves and a cinnamon stick

Put a teaspoonful of heather honey and a good slosh of Bushmill's into a mug. Add a squeeze or two of lemon juice to taste, and a couple of cloves and a bit of cinnamon stick if you've got them handy. Fill with recently boiled water and finish off with a slice of lemon. Drink straight away.

A traditional wassail

My mini-festival also marked the end of their year's toil and was a chance to drink, catch up with dear friends and celebrate the season now passed – even if it had been an absolute stinker. Naturally, it also served as an offering for fantastic bee health and warm weather with toasts of my home-brewed mead, an ancient brew that has witnessed a resurgence in the past few years. I intend to make a London version next year using the capital's honey; considering the specific characteristics of each different one it could be terrific.

The women and girls would wear bright garlands made from ivy and plastic flowers, hung with foam bees, whilst the chaps would stand around the honey-beer barrel or hog roast in the yard in their best tweeds and winter suits. Guests warmed their hands around old honey barrels in which I burnt old beehives and frames whose impregnated beeswax spat like firecrackers.

Farmer's wives would bring serious cakes and puddings, all served with thick heavy cream. I'm smiling with the memory as I write. Hell, it was a laugh. There would usually be a dance, a blessing and occasionally an open-air cinema showing beekeeping movies, projected from the back of my truck onto the wall of the nineteenth-century manor house by a friend's digital projector.

My plans to replicate this winter gathering in my London studio for urban landowners and clients, using the pulling power of bees and ping pong perhaps – the honey-packing table makes a great playing surface! – will have to wait until January. I also hope to make time for a New Year's Day ramble with all my fantastic bee volunteers. It's the perfect time for a romp around the capital as the streets will be empty, closed off for a swanky parade. My gang will be able to walk right down the centre of Piccadilly.

The aim is to visit all my London bee sites, plus some of the new ones that I'll be taking on next spring – we'll have some cake to keep us going, obviously, and at the end we'll pile into a favourite café in Old Compton Street for a good old fry-up. It'll be my way of saying thank-you for everyone's hard work and support throughout the year. All this is for after Christmas, though – just now, frankly, I'm wiped. I can't find the time or energy.

Instead I settle for a festive gathering of a few friends on the roof of Fortnum's the week before Christmas. The hives, I notice, failing to switch off from work, will need a good scrub in the spring, with a little Brasso on their copper and a touch-up on the gold leaf finials. They're now regularly visited by dignitaries attending the store and form part of a honey-tasting tour taken by their designer, Jonathan Miller. As this also attracts beekeepers, it's important the hives look their best for the new season.

But for now, a mince pie is eaten and a drop of whisky splashed on the hive, to give thanks to the bees for a great season and wish them a safe passage through the forthcoming barren months.

The bees, however, don't care about the décor, nor about the little party. I shouldn't do it, I know. But I can't resist a sneak peak beneath their heavily weighted oak roof. I see they're clustered around their fondant treat, nibbling away blissfully. Happy Christmas, my little Bees. See you in the New Year – when we start this ordeal again....

**Time for
new ideas**

Enough of my ramblings. Hopefully you've enjoyed hearing about my adventures, mishaps and business initiatives – no matter how deranged some of them may appear. I'd be delighted if you now feel inspired to keep bees yourself, or perhaps to share a hive if time and space are restricted.

Now is the time to act. There are numerous courses out there and, even if you don't end up taking the plunge and owning a hive, I hope you at least get to experience the critters close up and understand their magnificent and amazing life.

Also join me in my campaign to back the great resourceful hardy British-bred bee and all her amazing qualities and characteristics. I believe it is the best possible chance our bees have in the UK. Despite this huge upsurge in the craft and the demand for greater bee stocks, it's paramount wherever possible to use bees that have been specially bred to suit this country and climate, and not use foreign breeds.

This is going to mean further research and that in turn means investment. There are some fine individuals and organisations who are short of funding; they will be key to the future of these bees and I urge you to support them whenever possible. I'm encouraged to see the larger supermarkets supporting work on genetics now and I hope London will one day become fully pesticide free.

Worryingly, antibiotics were found in colossal proportions a few years ago in honey brought in from China. The consequences were a severe clampdown on honey imported into Europe, and many honey strains fail to make the grade thanks to stringent checks. I am told that honey is now listed at UK arrivals halls on a poster of not what to bring in, along with handguns and explosives – deadly stuff.

In many ways, this is a good thing. As less honey has become available from other parts of the world thanks to these import restrictions, people have turned to British honey (although so far producers in the UK haven't been able to satisfy the demand). It is also good news for people like David who import honey under the Fairtrade banner and are careful about the provenance of all their products.

So please, where possible use British bees, even if they are now mainly mongrels, and buy British honey. Retailers love the idea of stocking UK products such as honey, but have often been held back by availability. Fortunately, this appears to be changing as more UK beekeepers produce honey for sale and people begin to see British and local honey as something worth spending a little bit extra on. Long may this continue.

I also have a slightly off-the-wall idea – yes, another one – to set up a ploughing competition for vintage tractors in a London park. The turned soil could then be sown as a wild-flower meadow for bees with borage and sweet clovers. As yet I haven't floated this plan to others. My only concern would be the depth of the plough and whether there'd be a risk of disturbing any undiscovered Second World War incendiary devices.

Explosions aside, it will be an opportunity to bring my ageing Little Grey Fergie to London from Joe and Ann's shed in Merrington. Joe will be delighted to see the back of it. As for me, I'll fulfil my long-held dream of once again driving the tractor I sat on as a kid at Providence, my grandmother's smallholding where a love of bees was first instilled in me.

Bees are enchanting, hard-working and full of character, and their dedication to their task is just amazing. I can't imagine life without them. Do remember to tell your bees everything, just as I do ... you'll find it's very enlightening.

What all beekeepers should be doing:

- Make a plan for the following beekeeping year. Preferably in front of an open fire with a flagon of cider.
- Rest, enjoy this festive month and if you have sufficient honey look at selling it on of the many Christmas markets. Ensure you have trader's insurance for your products and public liability cover.

THANK YOU

As at the Oscars, darlings, I'd like to finish by thanking a few very important people, without whom none of this would have been possible. I promise not to blub.

David Wainwright in Wales: for letting me spend time with him at the pit face with his bees. These visits were the bucket of cold water in my face – the experiences that never let me become complacent about the amazing world of beekeeping. For getting me to think big and teaching me just about everything I know about commercial beekeeping. His knowledge of bees is extraordinary, just as his tales of early beekeeping exploits in London are inspiring. He is a true humble bee-master with the ability to work amazing hours and never tire.

Joe and Ann at Merrington Lane Farm: for their warmth, hospitality and fine tea and cakes, along with, of course, Joe's uniquely dry view on life. For tolerating all of my bee junk stored in their hay barn, along with my old Ferguson tractor, which I have great plans for in London.

David King (Kingy), the Gamekeeper: for his strength, survival instincts, true grit characteristics and love of bees. And for his unquestioning support in checking my hives at all hours of the day and night on remote moors and farmland. Also a big thank-you for those amazing gifts of pheasants, new potatoes, runner beans and sharp blackcurrants secretly deposited in my car. They were all, without fail, delicious and his devotion unrivalled.

Peter Kingsey, the remarkable bee breeder: for the woodland I rent, for his commitment to one particular strain of bee and for his gentle, Yoda-style approach to our craft. I mean this in the best possible way but I hope you never receive planning permission for the North London wood – it is the most brilliant urban haven for my bees and bee gear, that I would struggle to replicate as space becomes more of a premium.

Jill Mead, Ned's mum and my erstwhile partner in crime: for helping me launch my bee career and supporting me in those early years when others weren't willing to back our vision. They were incredibly happy times. We felt like true pioneers.

And of course my bees: for all their support and determination. They are my life and without them my existence now would be futile. Thank you, Bees. Your devotion has never been in question. Never.

You see, I told you it was like the Oscars....

ABOUT THE URBAN BEEKEEPER

Steve Benbow is at the heart of the urban beekeeping revolution. With zero experience he built his first hive on his tower-block rooftop ten years ago. Today, he runs hives across London's greatest landmarks, including both Tate galleries, Fortnum & Mason, and The National Gallery, as well as hives in Shropshire and Salisbury. He supplies honey to Fortnum & Mason, Harvey Nichols, the Savoy tearooms and Harrods, chefs including Gordon Ramsay and Marcus Wareing, and sells at local farmers markets and from his studio in Bermondsey. He also runs courses in beekeeping.

ACKNOWLEDGEMENTS

A special thanks also to:

Ian Belcher, Malcolm and Christine Finn, Sophie Campbell, Jennifer Cox, Tom and Lara Bean, Sarah Roberts, Jonathan Miller, Sam Rosen Nash, Sarah Hobbs, Barry the Ferret, Luke Lazenby, Emily Reed, Alecky Blythe, Bethan, Mandana, Markie B and Julius D and a huge thanks to Mary Chamberlain for editing at some extreme hours and for her husband's dangerous damson gin.

INDEX

Abel & Cole, 124
Aberystwyth, 20
acacia, 42
acacia honey, 42, 264
AFB (American Foul Brood), 56
Africa, 141 *see also* Cameroon; Zambia
Áine (volunteer), 100, 101-2
America, 133
American Foul Brood (AFB), 56
Amsterdam, 235
animal vandals, 245-8
Ann (author's friend), 30, 33, 41, 45, 256, 271, 273
antibiotics, 270
ants, 153, 246
aphids, 153, 154
apiary landlords, 255-6
artificial swarming, 85, 109-12, 116, 232
Ashes Hollow, 219
Asia, 141
Australia, 62, 133

badgers, 247
    honey, 161
Baklava French Toast, 204, 205-6

Bangla City hypermarket, 48, 49
Bangladesh, 213
Bangor University, 5, 186
barcodes, 206
bargains, 261
Barnes, 39, 245
Barry the Ferreter, 171, 259
bass, sea, 221
Battenberg cake, 31
Battersea, 151
Battersea Power Station, 124
BBC2, 204
Bean, Tom, 114-15, 126, 217, 220, 221, 222
    Honey with Mint and Peas recipe, 114, 115
Bea's of Bloomsbury, 15
bed sheet, white, 14, 107, 116
bee brush, 13-14
bee cab, 213-14
bee-eaters, 247
bee escape (clearer board), 144, 178-9
Beehaus hive, 231-2
bee inspectors, 82, 252
beekeeping associations, 29, 55, 63, 92, 213

beekeeping courses, 9, 21, 22, 230-1
beekeeping tasks, summary of
    for January, 22
    for February, 43
    for March, 63
    for April, 92-3
    for May, 116
    for June, 135-6
    for July, 162-3
    for August, 192
    for September, 215
    for October, 235
    for November, 253
    for December, 272
Beeny, Sarah, 104
bee space, 140
Beesplease, 25
bee stings, xii, 19, 40, 52, 54, 67, 103, 132, 151, 160-1
bee suit, 69, 130, 131 *see also* clothing
beeswax, 138, 139, 142, 181, 262-3
bee truck, 33, 58, 59-60, 61, 62, 65
bee vacuum, 107
Berkeley Hotel, 77-8
Berlin, 7
Bermondsey, 4, 15, 24, 31, 46, 56, 80, 101, 104, 114, 118, 125, 154, 180, 186, 195, 198, 199, 221, 241, 246, 252, 261, 262
Bernays, Lara, 222
    Lara's Heather Honey Harvest Cake, 222-3
Bethan (honey helper), 261
birds, 246-7
biscuits, honey, 174-5
blackthorn, 46, 95
Blair, Tony, 181

bluebells, 46, 148
blue tits, 246
Body Shop, 159, 162
borage, 124
borage honey, 124
Borough Market, 211, 212-13, 242
bottling, 260
Boycott, Rosie, 266
brace combs, 140-1, 229, 231
Bracket (Dame Hilda Bracket) (author's cat), 2, 62
breeders, 9, 29, 43, 73, 92, 128-9
breeding, 85-7
Bristol, 234
British Beekeepers' Association, 29, 41
Brixton, 152, 153
brood boxes, 81, 83, 85, 109, 138, 167, 170, 196
brood combs, 238
brood inspections, 81-4, 92, 135
brood rearing, 41, 51, 75
buckets, 10, 71, 180, 183
buddleia, 153, 195
Budgens, 206
bumblebees, 127-8, 195, 247
Butler Cottage, 2, 267

cakes, 30-2
    recipes, 15-17, 32, 222-3
California, 23
Camden Market, 212
Cameroon, 160
candles, beeswax, 208, 262-3
Capital Growth scheme, 251
Carluccio, Antonio: *Complete Mushroom Book*, 220
Carniolans, 106

Carr, William Broughton, 109
Cartland, Barbara, 2
caustic soda, 238
Cavendish Hotel, 28, 197
CBBC, 40
Central Science Laboratory, 55
ceps, 220
Chelsea Flower Show, 27
chemicals, 184, 201, 228, 239
chestnut mushrooms, 221
chestnut trees, 95, 195
chicken of the woods, 221
chicken wire, 247
children, 39-41
China, 124, 270
China Garden takeaway, Machynlleth, 19
Christmas markets, 261-3, 272
Chronic Paralysis, 149
cider vinegar, 57, 58
Clapham, 152
cleaning, 10, 22
clearer board (bee escape), 144, 178-9
clothing, 40, 52-4, 63, 151, 161, 188-9, 237-8 *see also* names of clothing items
Cloudy Wing Virus (Deformed Wing Virus), 54-5
clover, 133, 134, 148
clover honey, 134
CNN, 188
cold storage, 20, 194, 200-1, 202
Comb Honey with Munster Cheese Foam, 78-9
combining colonies, 196, 215
Coq d'Argent, 248
Cornwall, 233, 234
Crete, 168, 193-4

crocuses, 8, 41, 43
Crouch End, 206
crown boards/hive mats, 13 ,18, 215

dandelions, 46
Deformed Wing Virus (Cloudy Wing Virus), 54-5
Deli Downstairs, 204
deliveries, 242-3
Deptford, 113
disease, 28, 54-6, 63, 68, 72, 75, 82, 88, 98, 133, 140, 149, 197, 245, 252 *see also* Varroa
display stand, 264-5
Dockhead, 221
Dr Bernays' Honey Gar recipe, 57-8
Double Honey Ice Cream, 147-8
Doug (author's friend), 114, 150
doughnuts *see* Struffoli
'drawn-out' combs, 139
drones, 185-7, 219
dummy board, 12, 229-30
Dungeness, 181
dysentery, 38

East Anglia, 124
East London, 99, 102, 115-16, 120, 158-9, 204, 246
Eccles cake, 31
EFB (European Foul Brood), 56, 245
eggs, 82, 85, 86, 117
emergency queen/scrub queen, 86-7
entrance blocks/reducers, 19, 50, 75, 92, 145, 215, 248 *see also* mouse guards
equipment, beekeeping, 10-14, 22, 29, 35, 43, 63, 180, 231-2, 235 *see also* names of items

Eric (manager of Spitalfields Market), 101

Esther (former beekeeping pupil), 53, 103, 238

European Foul Brood (EFB), 56, 245

Euston, 109

excrement, bee, 57

extraction, honey, 2-3, 176, 180-2, 192, 194, 200, 202

extractors, 180, 182, 192

Fairtrade, 160, 271

feeders, 13, 71

feeding, 30, 47, 59, 63, 70-2, 93, 99, 135, 198, 228

Fernandez & Wells, 188

Flaming Hedgehogs, 221

fondant, 72, 77, 228, 235, 248, 269

Food and Environment Agency, 55

foraging trips, 117-18

Fortnum & Mason's, 6, 18, 27-8, 58, 102, 106-7, 126, 146-7, 186-7, 187-9, 190-1, 197, 200, 243, 246, 248, 260, 263, 264, 269

foundation wax, 138, 139

foxes, 47

frames, 10, 12, 29, 46, 81, 83, 84, 87, 107, 110, 112, 116, 138, 139-40, 141, 160, 200, 229, 238, 267

France, 75-6, 207

freezing of honeycombs, 194, 201, 202

fuel (for smokers), 89-90

fungi, 220-1

gaffer tape, 14, 69, 71, 108, 130, 131, 180

gamekeepers, 167, 171 *see also* King, David (Kingy)

ganja, 153

Gardner's store, Spitalfields, 100

Garnier Opera House, Paris, 11

gauntlets, 69

George (author's nephew), 174

Gerard (beekeeper in Paris), 40

gimp pins, 139

ginger cake, 31

ginseng honey, 234

Globe Theatre, 156

Glorious Twelfth (12th August), 166

gloves, 53, 69

goose wings/goose feather, 13-14, 159

Graves, David, 26, 226

Greece, 89-90

Greenwich, 120

Griffiths, Henry, 33, 58, 60, 62

grouse, 166-7

guard bees, 185

Guardian online, 256

Hackney, 99, 100, 120, 126, 134, 186, 191

Marshes, 125, 205

Hartley's jam factory, 198

Harvey Nichols, 233, 234, 235

hawthorn, 95

hay fever, 8, 76

Hayward, Richard, 211-12

head torch, 65-6

heater bees, 51, 250

heather, 165, 166, 167, 168, 169, 170, 219

heather beetle, 167, 219

heather honey, 125, 166, 170, 222, 224
　Lara's Heather Honey Harvest Cake,
　222-3
hedgehog mushrooms, 221
hefting, 18, 70, 228, 235, 258
Henley, 28
Herbert (Jack Russell), 30, 33, 34
Highlands, 171
Hinge (Dr Evadne Hinge) (author's
　cat), 2, 62
hive mats *see* crown boards/hive mats
hives
　amounts of honey produced by,
　　189-91
　assembling/building one's own, 10,
　　30, 35, 45, 171-2
　attacked by birds, 246-7
　bargains, 261
　Beehaus, 231-2
　bees maintain temperature in, 8, 51,
　　230, 250
　checking for signs of swarming,
　　106-7, 108, 162
　chemical treatment for Varroa, 184,
　　228
　choosing and purchasing, 10, 11,
　　12-13, 231-2, 261
　combining, 196
　having a spare, 88, 116
　hefting, 18, 70, 228, 235, 258
　increased activity at front of, 50, 117
　innovative design, 231-2
　inspecting/checking, 8-9, 18-19, 43,
　　81-4, 108, 116, 143, 192, 198
　invaded by bees, 196-7
　keeping off the ground, 38-9
　keeping records for, 14, 15, 112, 113

knocked over, 257
made from cedar wood, 11, 12
made from Douglas fir or other
　softwood, 11, 12
manipulation to prevent swarming,
　109-12
Modified Dadant, 12
monitoring honey stores in, 75
moving, 46, 58-62, 66, 67-70, 98-9,
　102, 113-14, 119, 122, 133,
　154-6, 165, 169-71
observation, 118-19, 146, 212, 213
opening for the first time, 52
organic treatment for Varroa, 184-5
overcrowding/lack of space in, 83,
　105, 106-7, 109
oxalic acid treatment for Varroa, 17-
　18, 22, 55, 184, 185
painting, 11-12
positioning, 63, 68-9, 123, 167, 173
preparing for winter, 194, 228-30,
　235, 237, 238
reducing entrance size, 145, 162,
　248, 229
refraining from opening, 50
removed and sterilised after loss of
　colony to disease, 54
sugar-shake testing for Varroa, 55,
　182-5, 192
traditional, 141, 160
vandalised by humans, 243-5
vandalised by wild creatures, 245-8
virtual, 41
and wasps, 145, 162, 248
WBC, 109
for hives at particular sites *see* names
　of locations

hive stands, 38-9
hive strap, 108, 130
hive tool, 11, 13, 84, 110, 196
Holloway, 266
honey
    acacia, 42, 264
    bees' stores of, 47, 59, 75, 83
    borage, 124
    clover, 134
    collecting in August, 175-80, 189,
        191, 192
    combs *see* honeycombs
    contamination by sugar syrup, 135
    dealing with honey flows in July,
        137-44, 150-1
    delivering, 242-3
    documenting Zambian harvesting of,
        159-62
    early, 80, 95-6, 114, 116
    exhibited at shows, 174, 232-3
    extracting, 2-3, 176, 180-2, 192,
        194, 200, 202
    Fairtrade, 160, 271
    at Fortnum & Mason's, 27, 187,
        188, 189, 190-1
    getting ready for, 81
    as gifts, 255-6, 261-2
    ginseng, 234
    and hay fever, 76
    heather, 125, 166, 170, 222, 224
    honeydew, 153, 154
    import restrictions, 270-1
    ivy, 227-8
    jars, 101, 163, 187, 200, 202, 207,
        260, 266
    labels/labelling, 101, 163, 187, 192,
        206, 261, 265

lime, 3, 7, 138
London, 4, 20, 25, 78, 95, 125, 143,
    180, 181, 187, 191, 201, 202, 212,
    234, 249, 264
manuka, 233-4
New York, 26
oilseed rape, 148
organic, 166
packing, 207-8, 240-1
recipes using, 15-17, 32, 57-8, 78-9,
    115, 127, 147-8, 174-5, 205-6,
    222-3, 249-50, 267-8
Salisbury Plain, 124-5
seeking to obtain subtle flavours in,
    46, 134
selling, 1, 101, 208-13, 215, 261-2,
    272
and spread of disease among bees, 56
stolen by bees, 196-7
stolen by wasps, 145
strange-coloured, 197-8
tasting, 77-8, 124-5, 266
thyme, 193
and tooth decay, 206-7
variable amounts from different
    hives, 189-91
see also honey boxes; honeycombs
honey badgers, 161
Honey Biscuits, 174-5
honey boxes
    bees eating honey in, 227
    checking available space on frames in,
        139
    ensuring enough available for use,
        162
    and Fortnum's hives, 190, 191, 192
    hauling into author's studio, 240-1

heavy weight when full, 143, 153, 199-200

marking with chalk to show apiary site, 264

number per hive, 12

preventing queen from laying in, 167

putting on, 81, 93, 102, 138

rapidly filled, 138

rodents in, 230

running short of, 137, 141

storage problem faced by author, 143

storage when empty, 10, 230

storage when full, 192

taking off, 143-4, 170-1, 175-80, 191, 237

theft of, 143

transporting, 20-1, 165

using trolleys to move, 264

warming before removing honey, 3, 240

Honey Cake recipes, 32, 222-3

honeycombs, 1, 20, 21, 46, 78-9, 138-9, 140-1, 147, 180, 194, 200-1, 201-2, 240, 241, 243, 263-4

    Comb Honey with Munster Cheese Foam, 78-9

    Double Honey Ice Cream, 147-8

honeydew, 15, 153-4

Honey Gar, 57-8

honey sac, 138

honey shows, 232-3, 235

Honey with Mint and Peas, 114, 115

hoods, 53

horse mushrooms, 221

Hospital for Tropical Diseases, 23

Hot Toddy, 267-8

Hughes, Simon, 211

Humphreys, Mark 'Chippy', 33

hygiene, 55-6

ice cream recipe, 147-8

icing-sugar shake, 55, 182-4, 192

import restrictions, 270-1

insecticides/pesticides, 6, 18, 149, 184

insurance, trader's, 272

Internet, 152

Italy, 10, 75-6, 207, 227, 258

ivy, 227-8

Jagger, Mick, 132

jam tarts 31

Jane (author's friend), 114

Jar of Honey from Mount Hybla, A (book), 27

jars

    honey, 101, 163, 187, 200, 202, 207, 260, 266

    for sugar-shake testing, 183

Jayne (author's sister), 32

    Honey Cake recipe, 32

Jermyn Street, 28, 188

Joe (author's friend), 30, 33-4, 45, 256, 271, 273

Josh (volunteer), 122, 124, 158

King, David (Kingy), 89, 171-3, 180, 218, 220, 224, 273-4

    Hot Toddy recipe, 267-8

Kingsey, Peter, 35-6, 36-7, 38, 50, 59, 73, 88, 89, 125, 130, 137, 274

Knappett, James, 77, 78

    Comb Honey with Munster Cheese Foam recipe, 78-9

knife, 181

labels/labelling, 101, 163, 187, 192, 206, 261, 265

Lara's Heather Honey Harvest Cake, 222-3

larvae, 77, 85

laurel, 244

lavender, 42

leaf blower, 144, 177-8

Leaton Estate, 259

lime honey, 3, 7, 138

lime trees, 3, 7, 46, 122, 195

ling, 165

Littlehampton, 162

London
    author's career as photographer in, 23
    author's decision to take his bees to, xiii, 4-7, 21
    author approached by Fortnum & Mason's, 27-8
    author takes over North London woodland site, 35-9
    author visits, 41-3
    author prepares for imminent arrival of bees in, 47
    author moves his bees to, 58-62, 65-7
    microclimate, 80, 84
    author's experience as beekeeper in, 4, 24-5, 28, 92, 95, 97-103, 104, 106-7, 109, 113-14, 115-16, 118-26, 134-5, 137-8, 142-3, 146-7, 150-1, 154-9, 168-9, 175-6, 185-6, 187-9, 190-1, 194-5, 197, 198, 199-201, 202-4, 206, 208-14, 225-7, 233-4, 237-9, 240-6, 247, 248, 251-2, 260-7, 268-9
    possible drone collection points in, 186-7
    author moves bees to Shropshire from, 168, 169-74
    wild food in, 221
    author brings bees and honey harvest back from Shropshire to, 224-5
    David Wainwright's experience of keeping bees in, 152-3
    brief references, 20, 29, 30, 34, 91, 166, 232, 270, 273
    see also names of sites in London

London, Mayor of, 251

London Bee Summit, 265-7

London Bridge Station, 157

London Electricity Board, 244

London Farmers' Markets, 210

London honey, 4, 20, 25, 78, 95, 125, 143, 180, 181, 187, 191, 201, 202, 212, 234, 249, 264

London Honey Company, 25, 101, 258, 263

London Honey Festival, 191

Long Meadow Junior School, xii

Long Mynd, 173
    Hike, 217-18

losses, 72-3, 74, 245

Ludlow food festival, 198

luggage-tag labels, 101, 206, 265

Luke (assistant to David Wainwright), 53

Luke (author's nephew), 248, 250

Machynlleth, 19

Malaysia, 23, 247

Maltby Street, 241

Mandana (volunteer), 150-2, 154, 155, 156, 157, 240, 267

Manhattan, 26
manuka honey, 233-4
Marble Arch, 203
marijuana, 152-3
markets, 208-13, 261-3, 265, 267, 272
  see also names of markets
Marks & Spencer, 202-3
mead, 268
Mead, Jill, 274
Merrington Lane Farm (farm of Joe and
  Ann), 271, 273 *see also* Ann; Joe
Mersea, 134
mesh bags, 102, 130, 155-6
mesh floors, 12-13, 69, 171, 237
mice, 34, 229, 230, 248 *see also* mouse
  guards
midday flight, 117, 136
Middle East, 234
Miller, Jonathan, 146-7, 188, 269
Miller feeder, 13, 71
Modified Dadant hive, 12
morel mushrooms, 221
More London, 42
mothballs, 201
mouse guards, 75, 92, 215, 228, 229,
  238, 247
mushrooms, 220-1

Nash (creative artist), 264-5
Nash, Sam Rosen, 126
  Summer Shortbread recipe, 126, 127
National Bee Unit, 55, 56
National Honey Show, 232, 235, 266
National Portrait Gallery, 146
National Theatre, 168
Natural Beekeeping Trust, 91
navigational behaviour, 118-19

Neal's Yard, 152
nectar, 42, 74, 80, 83, 96, 122, 135,
  138, 139, 142, 162, 165, 195, 227
Ned (author's son), 27, 39-40, 41, 199,
  224, 242
netting, builders', 123
New Year's Day ramble, 268-9
New York, 26, 121, 226
New Zealand, 133, 234, 252
North London site, 35-6, 37-8, 47, 61,
  62, 65-7, 89, 92, 95, 101-2, 116, 120,
  125, 134, 137, 154, 186, 201, 221,
  227, 247, 274
Nosema, 68
nucleus (nuc)/nucleus boxes, 14,
  110-11, 128-31, 135
nurse bees, 75

observation hives, 118-19, 146, 212, 213
oilseed rape, 6, 18, 37, 46, 125, 148-50
old pumping station site, 121-4, 158-9
Old Tannery (author's studio), 199,
  233, 240-2, 260-1, 264
Olive magazine, 147
Omlet (company), 231
open studio, 241-2
organic honey, 166
overalls, 53, 63, 188
overcrowding/lack of space, 83, 105,
  106-7, 109
overheating, 98-9
oxalic acid treatment, 17-18, 22, 55,
  184, 185, 239

packaging, 207-8, 240-1
painting hives, 11-12
paint-stripping gun, 181

Paris, 11, 121

Pembrokeshire, 73, 74

peppering, 85

pesticides/insecticides, 6, 18, 149, 184

Pestival, 213

pheromone, 86, 87, 111, 186

Philips, Roger, 221

Piccadilly, 27, 28, 269 *see also* Fortnum & Mason's

Pimlico, 210

pins
  gimp, 139
  stable, 69

Pitcairn Islands, 27

planning, 15, 22

Plumstead, 101, 244

Pole Cottage, 218-19

pollen, 7, 8, 30, 41, 42, 43, 50, 74, 75-7, 93, 107, 118

pollen dryers, 76

pollen sacs, 8, 30

pollen traps, 8, 76, 77

post, sending bees in, 73

Potters Field, 42

price guns, 261

propolis, 77

puffballs, 221

pumping station site, 121-4, 158-9

pyrethroid, 184

queen
  and artificial swarming, 109, 110, 111, 112
  bees cluster around, 250
  and brood inspections, 81-2, 108
  cared for by bees over winter months, 51, 230
  and David Wainwright's bees, 5, 21
  development affected by careless handling of queen cells, 73-4
  and drones, 186, 219
  and egg-laying, 30, 75, 82, 85, 117, 185, 218, 230
  failing, 85-6
  and honey boxes, 81, 167, 179
  and honey production, 189-90
  keeping separate during sugar-shake testing, 183
  looking after, 84-7
  making a new emergency queen, 86-7
  and new colony, 29, 128, 135, 189
  obtaining from breeders, 29, 73, 93, 128, 135, 136
  over-wintered, 73, 93
  raised by David King (Kingy), 172
  raised by Peter Kingsey, 38, 59
  recording age and breed of, 113
  replaced in third year, 59
  selecting to use as breeder, 59
  and space, 83, 85, 105, 109
  and swarming, 103, 105, 108, 109, 158-9

queen cages, 14, 111

queen cells, 73-4, 81, 83, 84-5, 92, 105, 108, 110, 111, 172

queen excluders, 81, 85, 93, 167

rats, 34, 230

recipes
  Baklava French Toast, 204, 205-6
  Comb Honey with Munster Cheese Foam, 78-9
  Dr Bernay's Honey Gar, 57-8

Double Honey Ice Cream, 147-8
Honey Biscuits, 174-5
Honey with Mint and Peas, 114, 115
Jayne's Honey Cake, 32
Kingy's Hot Toddy, 267-8
Lara's Heather Honey Harvest Cake, 222-3
Spiced London Honey Dressing, 249-50
Struffoli, 15-17
Summer Shortbread, 127
records/record books, 14, 15, 112-13, 116
ripeners, 260
Rio, 40, 121
rodents see mice; rats
Royal Festival Hall, 213, 252, 266

sage, wood, 181
sainfoin, 133
St John Bakery, Bermondsey, 31
Salisbury Plain, 133-4, 145, 221
Salisbury Plain honey, 124-5
schools, 40
    packs for use in, 41
screwdriver, 11
scrub queen/emergency queen, 86-7
sea lavender, 134
selling, 1, 101, 208-13, 215, 261-3, 272
'shook swarm' method, 56
shortbread, 126
    Summer Shortbread recipe, 127
Shrewsbury Flower Show, 174
Shropshire
    author's life in, xi-xiii, 2, 3, 4, 5-6, 26-7, 33-5, 45, 50

author moves bees to London from, 58-62
author visits, 148, 150
author's heather harvest expedition to, 168, 169-74
collecting bees from the moor, 217-20, 224-5
collecting and cooking fungi in, 220-1
wassail in, 267-8
brief references, 7, 30, 41, 80, 198, 247, 266
Shropshire Beekeepers' Honey Show Competition, 174
site checks, 257-8
Slovenia, 106
smell, bees' sense of, 36
smokers, 13, 89, 90-1, 153
smoking, 89 91
snow, 8, 237, 262
snowdrops, 7-8, 41, 43
Soho, 252
Sommariva del Bosco, 227
South Bank, 40, 191, 213
South Kensington, 262
South London, 106
space, 83, 85, 105, 109
Spain, 75-6, 207
Spiced London Honey Dressing, 249-50
spiders, xi, 10
Spitalfields Market, 100-1, 120, 207, 210
stable pins, 69
starvation, 72, 74, 83
Steele, Theo Fraser, 204
Steele, Sarah, 204
sterilising, 10, 22, 54, 238
sticky board, 17
stings see bee stings; wasp stings

Stiperstones range, 219
straw skeps, 141
stress, 98, 133
Stretton Springs water factory, 224
Struffoli, 15-17
sugar-shake testing, 55, 182-4, 192
sugar syrup, 13, 30, 47-9, 56, 63, 70-1, 72, 93, 99, 135, 198, 228
Summer Shortbread recipe, 127
sunshine, 68, 123
supers, 12, 85
super-seizure, 85
Surrey Quays, 152, 153
Sustain, 251
swarms/swarming, 9, 14, 83, 84-5, 91-2, 95, 97, 99, 102, 103-12, 115-16, 126, 158-9, 162, 172, 197, 232, 252
sycamore trees, 95
syrup *see* sugar syrup

Tarpaulin Mike, 48, 102, 155
tasting, 77-8, 124-5, 266
tasting box, 208
Tate galleries, 154-7, 198, 208
    Tate Britain, 157
    Tate Modern, 155-7, 225-6, 266
teeth, 206-7
television, 40, 188, 204
temperature
    in hives, 8, 51, 144, 230, 250
    outdoor, 8, 28, 50, 123, 250
Thames, River, 221
thyme, 42
thyme honey, 193
tools, 11, 35
torch, 65-6
Tower Bridge Primary School, 104
tractors, 33, 271, 273

Trading Standards, 192
training, 251 *see also* beekeeping courses
transportation
    author's experience of, 20-1, 58-62, 65-6, 97-9, 113-14, 133, 165-6, 170-1
    collecting new bees, 128-32
    major, 133
travel screens, 69
trees, 195 *see also* names of trees
Tregothnan Estate, 233
trout, 221-2
truck *see* bee truck
Turnbull, Bill, 260-1

utes (utility vehicles), 62

vandalism
    by humans, 243-5
    by wild creatures, 245-8
Varroa, 12, 17-18, 19, 22, 54-5, 63, 91, 182-5, 192, 228, 235, 238-40
vehicles, 133 *see also* bee cab; bee truck; tractors; utes (utility vehicles); Vespa
veils, 52, 53-4, 63
ventilation, 128, 129
Vespa, 60, 214
Viper's Bugloss, 133
virtual hive, 41
viruses, 149
Vo, Bea, 15
volunteers, 21, 37, 100, 122, 166, 240, 269 *see also* Áine; Josh; Mandana

Waggle Dance, 119
Wainwright, David
    author becomes acquainted with, 159, 162

author's beekeeping experiences with, 3-4, 19, 26-7, 133-4, 165, 166, 167, 168, 273

and author's plans for beekeeping in London, 5

and beekeeping equipment, 12-13, 39

breeding programme, 88

and cooking, 259

experience of keeping bees in London, 152-3

factory for processing honey, 20

and fuel for smokers, 89

gifts to author, 21, 256

holidays, 61-2

and honey for Fortnum's, 264

imports Fairtrade honey, 160, 271

lives in Wales, 5

owns weather unit, 97

supportive attitude to author, 159

visit to Zambia, 159-60

brief references, 66, 113, 114, 124, 148

Wales, 3, 5, 7, 20, 88, 160, 167, 187

Wareing, Marcus, 77

wasp guards, 145

wasps, 127, 145-7, 162, 247, 248

wasp stings, 146

wassail, 267-8

water, 113, 120, 129, 144, 162

water spray, 129

wax, 138, 139, 142, 181, 262-3

wax moths, 201-2

WBC cottage-style beehive, 109

weather
advantages of cold, 28
and appearance of queen cells, 84
and descent of bees through bee escape, 179
and disease, 72, 75
effect of poor spring weather on bees, 74-5
and foreign bees, 87
and heather, 166
and honey, 80, 81, 95, 134, 135, 180-1, 189, 196
impact on Christmas shoppers, 262
as key to good beekeeping season, 51
monitoring forecasts and reports, 51, 63, 81, 93, 97
and nectar, 96
and swarms, 106
and using a leaf blower for removing bees, 177, 178
see also snow

webcams, 6, 58, 106-7

Weiss, Mickael, 248, 250
Spiced London Honey Dressing recipe, 249-50

Wellcome Trust, 213, 214

Welsh cake, 31-2

Weybridge, 232

wheelbarrow, modified, 66

White City, 200

wimberries, 224-5

winter, getting ready for, 194, 228-30, 235, 237, 238

Women's Institute, 181

woodpeckers, green, 246-7

wood sage, 181

worker bees, 75, 86, 105, 138, 185

Yorkshire, 165-7

YouTube, 256

Zambia, 140, 159-62, 213

Hives on a barge
by Tower Bridge

Hives ought to be kept in good repair and well painted.

Pollen is mixed with honey to form bee bread which is fed to the grubs and to the older bees. It could be compared to bread and cheese as in bread for energy and cheese for protein and body building.

The Queen and the Drones stand from the same egg to form the 3 Groups in a colony leading the Queen. Royal jelly. All workers are female and come from fertilized eggs. The unfertilised eggs are hatched and produce Males.

The workers run the hive, unless the food pollen, make the wax for the comb. They make sure the hive is clean. At the end of the season the drones are banished and left outside to die.

B

There are many threats to bee colonies. A Fair instance the Varroa mite which drinks onto the bee. Most bee-keepers agree that it is possible the natural Varroa that kills the bee. Often it is one of around fifteen diseases it carries. It was described to me as 'bee HIV'.

Pesticides sprayed on plants and crops are picked up by the bees when out foraging. The bee itself will be harmed and worse still it will take the chemical cocktail back to the hive causing further danger to the colony.

The beekeepers hand pushed to hives cart.

MIEL
GARANTI
PUR
D'ABEILLES

E

MIEL
SURFIN

Miele di
alta montagna
delle Valle Maira

MIEL
GARANTI PUR
D'ABEILLES

GEWAARBORGD ZUIVERE
BIJEN-HONIG

Surprisingly London bees are more productive than their country relations. There are fewer threats to them in the city with fewer crops and little pesticide. Neither that the bees collect has little expense to the air there fore the city pollution does not affect the honey. The nectar is held so closely within the flower that the bee has to squeeze right in to get it.

Richard William Benbow
and Kate Benbow

n